Expert Consensus in Science

Anthony Jorm

Expert Consensus in Science

palgrave
macmillan

Anthony Jorm
Melbourne School of Population and Global Health
University of Melbourne
Carlton, VIC, Australia

This work was supported by the National Health and Medical Research Council (Investigator Grant APP1172889).

ISBN 978-981-97-9221-4   ISBN 978-981-97-9222-1   (eBook)
https://doi.org/10.1007/978-981-97-9222-1

This Palgrave Macmillan imprint is published by the registered company Springer Nature Singapore Pte Ltd.
The registered company address is: 152 Beach Road, #21-01/04 Gateway East, Singapore 189721, Singapore

If disposing of this product, please recycle the paper.

Preface

While working on this book, I have reflected on the pathway that led me to want to write it. Like many scientists, I once believed that science was all about what the data shows to be true and that expert consensus has little importance. After all, I thought, history has shown that the scientists of the past were often wrong in their views.

My beliefs about how science works can be traced back to my days as an undergraduate student in psychology at the University of Queensland in the early 1970s. My studies included a healthy dose of courses on research methods. The research methods I learned were all what we now refer to as "quantitative"—"qualitative" methods were never mentioned. If they had been, I certainly would not have thought of them as proper science. As a third-year student, I studied a book by Donald Campbell and Julian Stanley called *Experimental and Quasi-Experimental Designs for Research* (Campbell & Stanley, 1966). This book impressed me greatly for its clear thinking (and still does). It held up experimental methods as the strongest approach to causal inference and looked at the strengths and weaknesses of various experimental designs. In my early years as a researcher, I always aimed to use experimental methods wherever possible. Where experiments were not possible, I used other quantitative methods, such as longitudinal studies as the "next best".

In the mid-1980s, I moved into mental health research and learned a lot about the research methods that epidemiologists use, which again are all quantitative. I also learned about meta-analysis as a way of pooling data from a set to studies to get a more precise estimate of an effect. During the 1990s, there was a strong movement towards "evidence-based medicine", which I fully supported in its application to the treatment of mental disorders. One of the tools of the evidence-based medicine movement was the "hierarchy of evidence", which holds that a systematic review of randomized controlled trials is the strongest evidence, followed by a single well-designed randomized controlled trial. "Expert consensus" was typically relegated to the bottom of the hierarchy, which seemed appropriate to me.

The event that eventually led to my reconsideration of expert consensus was a conversation with my wife, Betty Kitchener, one evening when we were walking our dog. Betty was a registered nurse who had run first-aid courses for Red Cross. She also had a history of depression and, like me, shares a strong commitment to evidence-based practice. While on the dog walk, we had a conversation about why first-aid courses don't cover mental health crises, such as how to help a person who is suicidal, out of contact with reality or having a panic attack. After a few years of talking about the need for a public training course in the area and discussing what such a course would teach, we eventually developed the world's first Mental Health First Aid course in 2000.

In developing the course curriculum, we wanted to make it as evidence based as possible. The problem was that there was virtually no evidence on what a member of the public should do in a mental health first-aid situation, only on what professionals should do. When writing the course manual, we used what limited evidence we could find, sought advice from experts on various mental health problems about what a member of the public should do that might be helpful, and used common sense where there was nothing else to go on. However, we both felt this was not good enough. We asked ourselves what the techniques taught in conventional physical first-aid courses are based on and discovered the existence of international first-aid guidelines maintained by an organization called ILCOR. We decided that we needed to develop similar mental health first-aid guidelines.

If we were to use the hierarchy of evidence, we should try to carry out experimental studies of various mental health first-aid strategies. However, it is neither feasible nor ethical to randomly assign members of the public to different strategies. For example, we cannot instruct people to use one approach versus another in responding to a suicidal person in their social network. Thinking about what was feasible, we settled on expert consensus of professionals (clinicians and researchers) and people who had personal experience of a particular mental health problem or crisis. Using the Delphi method of assessing expert consensus, we developed a range of mental health first-aid guidelines on how to assist a person developing a mental health problem or in a crisis and rewrote the course content based on these guidelines. At the time of writing (2024), these expert consensus guidelines had informed the training of over seven million people in Mental Health First Aid courses in 29 countries.

Experience with Delphi studies greatly increased my respect for expert consensus as a source of knowledge, particularly as a way of gathering practice-based evidence from people with extensive practical experience. It also prompted me to notice how expert consensus underpinned so much of my work as a researcher, although often hidden from view. Even the "hierarchy of evidence" and, indeed, all the methodological tools of evidence-based medicine, I eventually realized, are based on expert consensus. Later, I read James Surowiecki's (2004) book on *The Wisdom of Crowds*, which provided a rationale for the use of expert consensus as a research method. Surowiecki argued that under certain conditions, groups of people with imperfect expertise could make valid judgements. Reading the primary evidence on the concept of "wisdom of crowds" showed me that while some consensus methods provide the conditions for good group decision-making (including the Delphi method that I had been using), others did not. I realized that psychological research on group decision-making held considerable promise for informing the processes of scientific consensus, which is a major theme of this book.

Another influence was the growing scientific and public discussion about climate change. On the one hand, climate scientists seemed to overwhelmingly believe that the climate is rapidly warming, and that human activity is the major contributor. On the other hand, there were some dissenting voices, and a common objection was that the consensus of climate scientists is not a good basis for determining the truth. After all, they argued, scientific consensus has sometimes been wrong in the past, citing examples like the Earth being seen as the centre of the universe or the continents as fixed in position. It seemed to me that there is a need to better inform scientists (and public commentators on climate change) about the importance of consensus of experts as an underpinning of science. I also thought that consensus methods used in the area of climate science could be improved.

A final motivation for writing the book is that I wanted to take the time to fully think through the ways in which expert consensus is involved in science. While I felt that my understanding had gradually improved over the years, I was aware that there were gaps, and I needed to make a concerted effort to put it all together. I am pleased to say that I enjoyed thinking things through and putting my ideas in writing.

Any book has to be written with a particular readership in mind. Because expert consensus is pervasive throughout all areas of science, I have attempted to write it to be relevant to working scientists and scientists-in-training from a broad range of disciplines, ranging from the physical to the social sciences. I have also aimed to write a book that will be of interest to practitioners and policymakers in areas where expert consensus plays an important role. This would include people concerned with evidence-based medicine, improving professional practice (e.g. the development of professional practice guidelines), and environmental policy (e.g. role of human activity in global warming). In particular, I have aimed to write a book which gives practical guidance on the best way to carry out a deliberative consensus process and critically evaluates the methodologies for achieving this.

Throughout the book, I have used a broad range of case examples to illustrate my general points. My hope is that these case examples bring the general conclusions to life (but the reader who wants a quick overview can skip them and still follow my arguments). However, I am aware that the case examples are largely from scientific activities in high-income countries, particularly English-speaking ones. This reflects the global inequalities of scientific activity and does not imply a lesser importance to scientific consensus from other parts of the world.

Given the size of the topic, this book is arguably short. This is deliberate. As a working scientist myself, I know that time is too limited to read everything I want to. I appreciate it when others can write clearly and concisely, and always aim to do this myself. I must admit to occasional frustration at the long-windedness and opacity of some of the works I needed to read when writing this book.

I originally planned to begin writing this book in 2019. At the start of that year, I officially "retired" from my position as a professor at the University of Melbourne and took on an unpaid emeritus professor role. My retirement was only nominal. After more than three decades of holding competitive National Health and Medical Research Council Fellowships, I thought I had had a pretty good run and needed to move out of the competition for future fellowship funding so that younger researchers could have a turn. I had a healthy superannuation balance that meant that I could live quite comfortably without a fellowship salary and continue doing my research as before. One of my aims during my "retirement" was to write this book. However, I was so engaged in various research projects that it got put on the back burner. What finally prompted me to start writing was a conversation I had with my niece Christine Jorm in 2022 where I told her about the idea for the book. She was very enthusiastic and told me, "You really must write that book!" As Christine has had wide experience as a medical specialist, biomedical researcher, health sociologist and health administrator, I was encouraged by her interest and began writing seriously in 2023. Christine also provided me with comments and suggestions for improvements on the entire first draft.

There were a number of other people who greatly assisted me with the book. My principal thanks go to my wife Betty Kitchener. As Betty and I have done many Delphi studies together, she was well aware of the issues involved in establishing an expert consensus. During our regular evening walks, she was a constructive sounding board about my thoughts when working on a chapter, leading to some important changes in the organization of the work. She also gave helpful comments on the first draft of every chapter after it was written. There have also been a number of anonymous reviewers who provided comments which challenged me and extended my thinking. Although I don't know who they are, I wish to thank them for their contribution.

Carlton, VIC, Australia   Anthony Jorm

References

Campbell, D. T., & Stanley, J. C. (1966). *Experimental and quasi-experimental designs for research*. Rand McNally. https://books.google.com.au/books?id=kVtqAAAAMAAJ

Surowiecki, J. (2004). *The wisdom of crowds: Why the many are smarter than the few*. Doubleday.

Contents

List of Figures

List of Tables

1

The Controversy over Expert Consensus in Science

A book on "expert consensus in science" should begin by defining what is meant by the term. However, the answer is complex and requires consideration of the many uses that consensus processes have in science. These complexities are discussed in detail in subsequent chapters of this book. However, in the interim, I provide a basic definition which will suffice for now: "expert consensus in science" is a high level of agreement among scientists with relevant expertise about a specific scientific claim, methodology or science-based practice or policy.

In the not-too-distant past, the topic of expert consensus in science would only have been of interest to scientists and scholars in the history, philosophy and sociology of science. However, there is now a much broader interest in the topic from both scientists and non-scientists alike. A major reason for this is that expert consensus is increasingly used to guide national policies based on scientific findings, but some of these policies may be at odds with people's pre-existing beliefs and values and they understand little about scientific consensus.

The pre-eminent contemporary example is the role of human activity in global warming. The United Nations Environment Programme and the World Meteorological Organization established the Intergovernmental

© The Author(s) 2025
A. Jorm, *Expert Consensus in Science*, https://doi.org/10.1007/978-981-97-9222-1_1

Panel on Climate Change (IPCC) in 1988. The IPCC uses an expert consensus process, with reports going through a series of steps, including governments and observer organizations nominating experts as potential authors, drafting of reports by the authors which are reviewed by a large number of experts, revision based on feedback, and approval by all governments in the United Nations of the final documents (Intergovernmental Panel on Climate Change, 2023). The IPCC produced its first report in 1990, with subsequent reports in 1995, 2001, 2007, 2014 and 2023. Over successive reports, the consensus conclusions about the role of human activity in global warming have become progressively stronger. In its latest report, the IPCC (Intergovernmental Panel on Climate Change, 2023) concluded with "high confidence":

> Human activities, principally through emissions of greenhouse gases, have unequivocally caused global warming, with global surface temperature reaching 1.1°C above 1850–1900 in 2011–2020. Global greenhouse gas emissions have continued to increase, with unequal historical and ongoing contributions arising from unsustainable energy use, land use and land-use change, lifestyles and patterns of consumption and production across regions, between and within countries, and among individuals. (Section A.1)

The report also concluded with "high confidence":

> Widespread and rapid changes in the atmosphere, ocean, cryosphere and biosphere have occurred. Human-caused climate change is already affecting many weather and climate extremes in every region across the globe. This has led to widespread adverse impacts and related losses and damages to nature and people. (Section A.2)

Further contributing to the public interest in the issue, climate change is constantly in media headlines, because of extreme weather events such as record high temperatures, massive wildfires and melting of ancient icefields. Such events are often accompanied by comments on the "scientific consensus" attributed to the IPCC.

People in prominent policy positions are now expected to take a stance on the issue of climate change, even if they are not trained as scientists.

In 2023, the president of the World Bank stepped down from his position after criticism about his lack of action over the issue. As reported in the press at the time (Civillini, 2023):

> World Bank president David Malpass will step down from his post in June, nearly a year before his term is due to expire. Malpass received strong criticism over the bank's commitment to climate action and over his personal views on climate change. He had been under increasing pressure since last September, when he refused to publicly accept that burning fossil fuels is warming the planet. Malpass was asked during an event on the sidelines of the UN general assembly whether he agreed with the scientific consensus on climate change. The World Bank chief repeatedly dodged the question, to heckling from the audience, before eventually responding "I'm not a scientist". (Sentences 1–5)

Adding confusion in the minds of members of the public, there have also been dissenting groups which have promoted their own consensus. The Nongovernmental International Panel on Climate Change (Nongovernmental International Panel on Climate Change, 2023) has concluded that natural causes rather than human activity are the dominant cause of climate change, and there is a World Climate Declaration, which is a petition signed by a varied group of scientists who dispute a number of the IPCC's conclusions (Global Climate Intelligence Group, 2022).

Similar to the situation with climate change, expert consensus processes have been used to guide policies on other issues where there has been strong public interest. One of these is the safety of food produced from genetically modified (GM) crops. Consensus statements supporting GM food safety have been issued by a wide range of organizations, including the American Medical Association, the American Association for the Advancement of Science, Food Standards Australia New Zealand, the French Academy of Science, the Royal Society of Medicine, the European Commission, the Union of German Academies of Sciences and Humanities and a number of national Academies of Sciences (Norero, 2022). Nevertheless, as with climate change, there are scientific dissenters. A joint statement signed by over 300 researchers who disputed these consensus statements has been published under the title *No scientific consensus on GMO safety* (Hilbeck et al., 2015). The authors concluded:

> the scarcity and contradictory nature of the scientific evidence published to date prevents conclusive claims of safety, or lack of safety of GMOs. Claims of consensus on the safety of GMOs are not supported by an objective analysis of the refereed literature. (p. 1)

Such conflicting statements about the scientific consensus can only contribute to increasing scepticism about consensus processes in general.

A third issue of considerable public interest is vaccination of infants and children. In this case, there is a strong expert consensus in favour of vaccination (Dornbusch et al., 2017), and the challenge has come from sceptical parents and lay activists rather than dissenting scientists. One source of opposition to vaccination is the belief that it might increase risk of autism or other neurological disorders (Stolle et al., 2020). This belief was bolstered by a study by Andrew Wakefield that was published in *The Lancet* in 1998. Although this paper has been debunked and retracted, and multiple studies have failed to find any association (Taylor et al., 2014), the alleged link still has lay proponents. Another source of opposition to the consensus on vaccination is the belief that vaccines contain mercury, a known neurotoxin, which has been championed by US environmental lawyer and politician Robert F Kennedy Jr (Jarry, 2021). Again, the belief has persisted despite the evidence against it (Taylor et al., 2014).

Major factors in the rejection of the expert consensus on vaccination are a mistrust of science and of the pharmaceutical industry, and a focus on personal liberty and parental rights (Carpiano et al., 2023; Stolle et al., 2020; Sturgis et al., 2021). Such influences present a major challenge to the public acceptance of expert consensus, as they raise the issue of whether the public should trust the experts or attempt to evaluate the evidence for themselves, personally weighing up the pros and cons of various courses of action.

1.1 The Consensus Sceptics

Given that members of the public and scientists sometimes have views that differ from the expert consensus on the above issues, there has understandably been questioning of the role of consensus in science more generally. There are a number of prominent people who have taken such a contrarian position. One of these is Michael Crichton, an American author and filmmaker whose works often dealt with science-related themes. Probably his best-known work was the novel *Jurassic Park*, which was subsequently made into a box-office hit movie by Stephen Spielberg. Michael Crichton was not a scientist, but had firm views on a number of scientific matters, including the science behind global warming. In 2003, Crichton gave a public lecture at the California Institute of Technology (Caltech) as part of its Michelin Distinguished Visitors Lecture Series. This lecture series was established to foster creative interaction between the arts and scientists. The intriguing title Crichton chose for his lecture was "Aliens Cause Global Warming" (Crichton, 2003). In the lecture, Crichton covered a number of topics that he argued were "bad science", including the search for extraterrestrial intelligence, the possibility of a Nuclear Winter following a nuclear war, predictions about overpopulation, in addition to human influence on global warming. He was particularly critical of statements about what "the scientific consensus" said, which he saw as underpinning these and other examples of bad science. He stated:

> Let's be clear: The work of science has nothing whatever to do with consensus. Consensus is the business of politics. Science, on the contrary, requires only one investigator who happens to be right, which means that he or she has results that are verifiable by reference to the real world. In science consensus is irrelevant. What is relevant is reproducible results. The greatest scientists in history are great precisely because they broke with the consensus.

> There is no such thing as consensus science. If it's consensus, it isn't science. If it's science, it isn't consensus. Period.

In addition, let me remind you that the track record of the consensus is nothing to be proud of. (p. 5)

Crichton's views on consensus in science are not unusual. Another example comes from an article in the *Financial Times* by John Kay (2007), an economics and business commentator, who wrote:

> We do not say that there is a consensus over the second law of thermodynamics, a consensus that Paris is south of London or that two and two are four. We say that these things are the way things are….Numbers are critical to democracy, but science is not a democracy….Science is a matter of evidence, not what a majority of scientists think…Statements about the world derive from their value and arguments that support them, not from the status and qualifications of the people who assert them…The notion of a monolithic "science", meaning what scientists say, is pernicious and the notion of "scientific consensus" actively so. (Paras. 4–9)

Another example of a sceptic about scientific consensus is former Australian prime minister Tony Abbott. In a speech to the Global Warming Policy Foundation in the United Kingdom in 2017, he argued that industries and living standards were being sacrificed in order to reduce CO_2 emissions, to little benefit. He dismissed the argument that a large majority of scientists believed that human activity was contributing to climate change, stating that "the claim that 99 per cent of scientists believe" is "as if scientific truth is determined by votes rather than facts" (Yaxley, 2017).

The doubts that these laypeople have about expert consensus are understandable, because the sort of science they learned about in high school and the scientific "breakthroughs" they hear about regularly in the press are based on major research studies rather than a process of coming to a consensus. As Curry and Webster (2013) have noted:

> With genuinely well-established scientific theories, 'consensus' is not discussed and the concept of consensus is arguably irrelevant. For example, there is no point in discussing a consensus that the Earth orbits the sun, or that the hydrogen molecule has less mass than the nitrogen molecule. (p. 3)

While it may be easy to dismiss the views of prominent people who lack scientific expertise, similar statements questioning the role of consensus in science have appeared in articles published in reputable scientific journals. In a 2009 editorial in the journal *Molecular Imaging and Biology*, the editor-in-chief, Jorge Barrio (2009), quoted Michael Crichton approvingly, adding:

> It is indeed hard to disagree with Mr. Crichton. The historical track record of scientific consensus is nothing but dismal. Many examples can be cited, but there are some classical ones. Nicholas Copernicus and his follower, Galileo Galilei, experienced the effects of consensus when they advanced theories that planet Earth was not the center of the Universe. The sixteenth and seventeenth centuries were not the right time to go against established dogmas.

> Today, the methods for exacting consensus have changed but the result could be the same: The death of the spirit. The use and abuse of "consensus science" is at least partially responsible for the current crisis in the scientific and medical peer review system. (p. 1)

Similarly, in a 2019 article in the journal *Dose-Response*, Yehoshua Socol et al. (2019) argued:

> Appealing to scientific consensus is an adequate tool in policy-making and public debate. However, appealing to consensus often occurs in scientific discussion itself, which is absolutely unacceptable... Consensus is not an argument in scientific discussion; only experimental evidence matters. There are examples of decades-long scientific consensus on erroneous hypotheses. (pp. 1 and 4)

More recently, in 2021, Kamran Abbasi, the editor-in-chief of the prestigious British medical journal *BMJ*, asked, "Does consensus even matter when it's the evidence that should matter?" (Abbasi, 2021). Commenting on medical practice guidelines, he stated:

> Yet many guidelines are based on consensus and sold to us in such a way that we might assume the authority of the assembled experts to be greater

than the accumulated evidence. New research analysing US guidelines in cardiology and oncology instead finds that consensus based guidelines are more likely to make discordant and inappropriate recommendations relative to the evidence base. (p. 1)

Such scepticism about the value of expert consensus is by no means universal among scientists, but it does represent a common view. Indeed, it is sometimes embedded in frameworks for evaluating the quality of evidence. An example comes from the work of JBI, an organization that systematically reviews the research evidence on health in order to improve the quality of healthcare. JBI has a scheme for rating the Level of Evidence for effectiveness of health interventions, with higher levels seen as superior to lower ones (JBI, 2013). The levels are:

- Level 1: Experimental designs
- Level 2: Quasi-experimental designs
- Level 3: Observational–analytic designs
- Level 4: Observational–descriptive studies
- Level 5: Expert opinion and bench research

Expert opinion is in the lowest level, Level 5, which is itself divided into three sub-levels:

- Level 5.a: Systematic review of expert opinion
- Level 5.b: Expert consensus
- Level 5.c: Bench research/single expert opinion

The JBI framework is not unique in this respect. There are a number of similar frameworks for rating Levels of Evidence which also give a low score to evidence based on expert consensus.

Other critics accept that consensus does play a role in science, but argue that the consensus on issues like climate change is biased because there is pressure on scientists to conform or the consensus is "manufactured" by excluding experts who have dissenting views. An example of such views comes from climate scientist Judith Curry (2022), who accepts that global temperatures have been rising and that CO_2 emissions by humans will act to warm the planet. However, she argues that the IPCC consensus process exaggerates the risk and brushes over the uncertainties in the evidence for political purposes. She states:

> What we do object to is the idea of a manufactured consensus for political purposes. This is not a natural scientific consensus that has emerged over a long time. It's a manufactured consensus of scientists at the request of policy makers, which has been too narrowly framed. There's too much politics in it. And that's what I object to and there's a number of other scientists that object to this as well. And we've also been critical of the behaviour of some of the more politically active scientists who are exaggerating the truth in the interests of a good story or political objectives. (para. 2)

The pressure on scientists to conform or compromise when producing a consensus statement has also been noted by Daniel Sarewitz (2011), a scientist who has a participant in the production of a consensus document on *Geoengineering: A National Strategic Plan for Research on Climate Remediation*. In a commentary in the journal *Nature* in which he reflected on his participation, Sarewitz concluded that: "The discussions that craft expert consensus... have more in common with politics than science" (p. 7). He went on to state:

> The very idea that science best expresses its authority through consensus statements is at odds with a vibrant scientific enterprise. Consensus is for textbooks; real science depends for its progress on continual challenges to the current state of always-imperfect knowledge. Science would provide better value to politics if it articulated the broadest set of plausible interpretations, options and perspectives, imagined by the best experts, rather than forcing convergence to an allegedly unified voice. (p. 7)

These concerns about pressure to conform and manufactured consensus raise the possibility that expert consensus processes may sometimes be poorly done, leading to doubts about the conclusions in consensus statements.

1.2   The Contribution of This Book

Given these controversies, this book aims to examine the role of expert consensus in science. It shows how consensus processes pervade science, being important to establishing what is regarded as scientific truth, developing science-based guidance on professional practice and public policy, and agreeing on what research methodologies are sound. There are a range of methods that scientists have used to establish a consensus, but these could be improved using psychological research into how groups of individuals make good judgements. It is also argued that if we are to persuade the public to adopt science-informed views on issues like climate change, there needs to be greater education about the importance of consensus in science. The specific contributions of each chapter to these conclusions are as follows:

Chapter 2 argues that contrary to the views of the consensus sceptics, consensus processes pervade science. Using examples from a range of scientific areas, it shows how consensus is involved in generating ideas and setting priorities, assessing funding applications and distributing access to resources, selecting methods to use in a research project, publication of scientific findings, reviewing the published literature and drawing conclusions on facts. Taking a specific example of a research project, it identifies ten points during the research where consensus processes were involved.

Chapter 3 examines the most controversial area where consensus is involved—establishing scientific truths. It reviews the range of positions taken by philosophers, historians and sociologists of science, from those who take a strongly positive position through to those who are largely negative about the role of consensus in establishing truth. Despite the varying views, there is some agreement from these writers about the conditions under which consensus is more likely to indicate truth: (1) The consensus needs to be rational, empirical and critically examined. (2) The group coming to the consensus needs to be diverse. (3) The group needs to be open-minded, and there is no coercion of dissenters. (4) The group needs to be sufficiently large to get reliable results. It concludes that consensus can be a strong indicator of truth under certain conditions.

Chapter 4 proposes two contrasting processes by which scientists come to a consensus, which are labelled as "spontaneous" and "deliberative". The spontaneous process involves a consensus that develops rapidly and spontaneously among experts in an area. It is more likely to be seen with scientific questions that involve simpler causality and strong associations between variables. The development of a consensus is hidden from view and may lead an outside observer to think that the scientific facts emerge directly from the evidence. The deliberative process, on the other hand, is much slower, and it may take decades to come to a consensus. It is more typical with complex scientific questions where the evidence is extensive and involves multiple disciplines and methodologies. It involves formal methods to develop the consensus, such as consensus conferences, expert working groups set up by international scientific organizations, Delphi consensus studies of expert opinion and formal votes by groups of acknowledged experts. Deliberative consensus is becoming more important as scientists deal with increasingly complex problems in areas of global importance where coordinated action is required.

Chapter 5 looks at how deliberative consensus is often used to making evidence-based recommendations to guide professional practice and public policy. When used for such purposes, deliberative consensus involves considerations additional to what the scientific evidence shows, in particular value judgements about various courses of action. The evidence-based medicine movement is examined as an influential example of the use of deliberative consensus to guide medical practice. However, consensus processes have also been used to develop guidelines and position statements on practice and policy in other areas, with the work of the Intergovernmental Panel on Climate Change an important example.

Chapter 6 discusses how consensus processes are used by scientists to agree on what research methodologies in their field are sound. Many methodological innovations are accepted by a spontaneous consensus. This is more likely where there is no existing method or existing methods can be improved. With such methods, the level of innovation is high and can be provided by an individual scientist or a small team. Other methodological innovations achieve consensus through a deliberative process. This more commonly occurs where there are existing methods, but these need standardization or infrastructure for dissemination. It requires coordination of efforts across a larger number of scientists and may require creating a new organization to support the dissemination of the methodology.

Chapter 7 looks at how expert consensus processes specify who is an "expert" and what constitutes "consensus". There are a number of attributes that have been used to specify who is a scientific expert, including professional qualifications and work experience, membership of scientific or professional organizations, peer-reviewed publications, specialist conference attendance and nomination by other experts. A common factor across these attributes is acknowledgement of expertise by peers. Where consensus has to be reached on matters of values as well as scientific questions, the values of all interested parties must be considered, which may include the general public, cultural experts or consumer advocates as well as scientists. When consensus occurs spontaneously, there is no formal process to ascertain agreement, but there are indicators that it has occurred, such as a high rate of positive citations and incorporation in textbooks. With deliberative consensus the level of agreement among experts is quantified. However, what level of agreement is required for "consensus" depends on the purpose, with a higher level needed for establishing likely scientific truth than for defining concepts and standardizing measures.

Chapter 8 describes the range of methods that have been used for determining deliberative consensus. These are Delphi studies, the nominal group technique, surveys of experts, systematic analysis of conclusions in the peer-reviewed literature, consensus conferences and expert working groups. There are also emerging methods which are not yet in common use: scientific citation networks, prediction markets and artificial intelligence.

Chapter 9 reviews research from psychological science on the conditions under which groups make optimal judgements, a subject area often called "wisdom of crowds". It concludes that good judgements are more likely when the members of a group are selected for expertise, there is cognitive diversity about the members, they make independent judgements which are then aggregated, and there is opportunity for sharing information and discussion. When the methods that scientists use to establish deliberative consensus are evaluated against these conditions, none meet them all, but some (Delphi studies and nominal group technique) are better than others.

Chapter 10 proposes the need for a new area of research on the "wisdom of scientific crowds", which investigates how groups of scientists make optimal judgements using tasks more typical of those that face scientists. It reviews seven realistic scientific judgement tasks that could be used for this purpose.

Chapter 11 looks at how scientific conclusions should be communicated to the public. It reviews evidence that communicating the scientific consensus on an issue can change public beliefs. However, some people reject the scientific consensus on issues like climate change and vaccine safety, because they do not trust scientists, basing this mistrust on the clash between their own values and the values inherent in the scientists' consensus, and they may overestimate their own understanding of very technical areas. A possible way forward is to educate people about the important role of consensus in science more generally, rather than focus solely on consensus messages about specific scientific issues. This would require a greater understanding of the role of consensus in science at all levels of science education from high school through to specialist postgraduate training.

References

Abbasi, K. (2021). The dangers in policy and practice of following the consensus. *BMJ, 375*, n2885. https://doi.org/10.1136/bmj.n2885

Barrio, J. R. (2009). Consensus science and the peer review. *Molecular Imaging and Biology, 11*(5), 293. https://doi.org/10.1007/s11307-009-0233-0

Carpiano, R. M., Callaghan, T., DiResta, R., Brewer, N. T., Clinton, C., Galvani, A. P., et al. (2023). Confronting the evolution and expansion of anti-vaccine activism in the USA in the COVID-19 era. *Lancet, 401*(10380), 967–970. https://doi.org/10.1016/S0140-6736(23)00136-8

Civillini, M. (2023, February 16). World Bank chief to step down early after climate controversy. *Climate Home News*. https://www.climatechangenews.com/2023/02/16/world-bank-chief-steps-down-climate-controversy/#:~:text=World%20Bank%20president%20David%20Malpass,personal%20views%20on%20climate%20change

Crichton, M. (2003). Aliens cause global warming. In California Institute of Technology (Ed.), *Caltech Michelin Lecture*. Pasadena.

Curry, J. (2022, October 5). There's no climate emergency. *BizNews*. https://www.biznews.com/global-citizen/2022/10/05/climate-change-2

Curry, J. A., & Webster, P. J. (2013). Climate change: No consensus on consensus. *CAB Reviews, 8*(001). https://doi.org/10.1079/PAVSNNR20138001

Dornbusch, H. J., Hadjipanayis, A., Del Torso, S., Mercier, J. C., Wyder, C., Schrier, L., et al. (2017). We strongly support childhood immunisation-statement from the European Academy of Paediatrics (EAP). *European Journal of Pediatrics, 176*(5), 679–680. https://doi.org/10.1007/s00431-017-2885-0

Global Climate Intelligence Group. (2022). *World climate declaration: There is no climate emergency*. www.clintel.org

Hilbeck, A. B. R., Defarge, N., Steinbrecher, R., Szekacs, A., Wickson, F., Antoniou, M., Bereano, P. L., Clark, E. A., Hansen, M., Novotony, E., Heinemann, J., Meyer, H., Shiva, V., & Wynne, B. (2015). No scientific consensus on GMO safety. *Environmental Sciences Europe, 27*, 4. https://doi.org/10.1186/s12302-014-0034-1

Intergovernmental Panel on Climate Change. (2023). *Climate change 2023: Synthesis report. Contribution of Working Groups I, II and III to the sixth assessment report of the Intergovernmental Panel on Climate Change*. IPCC.

Jarry, J. (2021, April 16). The anti-vaccine propaganda of Robert F. Kennedy, Jr. *Office for Science and Society (OSS) Weekly Newsletter*. https://www.mcgill.ca/oss/article/covid-19-health-pseudoscience/anti-vaccine-propaganda-robert-f-kennedy-jr

JBI. (2013). *JBI levels of evidence*. https://jbi.global/sites/default/files/2019-05/JBI-Levels-of-evidence_2014_0.pdf

Kay, J. (2007, October 10). Science is the pursuit of the truth, not consensus. *Financial Times*. https://www.ft.com/content/c49c8472-767b-11dc-ad83-0000779fd2ac

Nongovernmental International Panel on Climate Change. (2023). *NIPCC Nongovernmental International Panel on Climate Change*. The Heartland Institute. Retrieved February 21, 2023, from http://climatechangereconsidered.org/

Norero, D. (2022). *GMO 25-year safety endorsement: 280 science institutions, more than 3,000 studies*. Genetic Literacy Project. https://geneticliteracyproject.org/2022/01/21/gmo-20-year-safety-endorsement-280-science-institutions-more-3000-studies/

Sarewitz, D. (2011). The voice of science: Let's agree to disagree. *Nature, 478*(7367), 7. https://doi.org/10.1038/478007a

Socol, Y., Shaki, Y. Y., & Yanovskiy, M. (2019). Interests, bias, and consensus in science and regulation. *Dose-Response, 17*(2), 1559325819853669. https://doi.org/10.1177/1559325819853669

Stolle, L. B., Nalamasu, R., Pergolizzi, J. V., Jr., Varrassi, G., Magnusson, P., LeQuang, J., et al. (2020). Fact vs fallacy: The anti-vaccine discussion reloaded. *Advances in Therapy, 37*(11), 4481–4490. https://doi.org/10.1007/s12325-020-01502-y

Sturgis, P., Brunton-Smith, I., & Jackson, J. (2021). Trust in science, social consensus and vaccine confidence. *Nature Human Behaviour, 5*(11), 1528–1534. https://doi.org/10.1038/s41562-021-01115-7

Taylor, L. E., Swerdfeger, A. L., & Eslick, G. D. (2014). Vaccines are not associated with autism: An evidence-based meta-analysis of case-control and cohort studies. *Vaccine, 32*(29), 3623–3629. https://doi.org/10.1016/j.vaccine.2014.04.085

Yaxley, A. (2017, February 19). *Tony Abbott says climate change action is like trying to 'appease the volcano gods'.* https://www.abc.net.au/news/2017-10-10/tony-abbott-says-action-on-climate-change-is-like-killing-goats/9033090

2

Consensus Pervades Scientific Processes

The previous chapter quoted Michael Crichton's (2003, p. 5) claim that "The work of science has nothing whatever to do with consensus".

The present chapter argues that this view is misguided, and that expert consensus pervades scientific processes. The chapter does this by taking the reader on a tour through the various phases of scientific research, starting with generating scientific ideas and continuing through to drawing conclusions about scientific facts, showing how consensus processes are involved in each phase. To illustrate how consensus is involved at various phases, case examples are presented from a wide range of scientific disciplines, illustrating that consensus processes are important across the broad spectrum of science. These examples are used to illustrate what actually happens in science and should not necessarily be seen as best practice for how consensus should be established. As argued later in this book, scientific consensus processes can be improved.

This tour through the phases is necessarily piecemeal and may not give the reader a holistic picture of how multiple consensus processes are involved in a single research project. To complement the phase-by-phase approach, later in the chapter I also provide a detailed example from my own research, showing how consensus processes played an essential role

© The Author(s) 2025
A. Jorm, *Expert Consensus in Science*, https://doi.org/10.1007/978-981-97-9222-1_2

at many points in a single project which was concerned with what parents of teenagers can do to reduce the risk of anxiety problems and depression in their children.

2.1   Use of Consensus to Generate Ideas and Set Priorities

Science begins with generating scientific ideas (hypotheses, theories). Sometimes this will be the work of individual scientists. However, these ideas are often floated with colleagues, who provide critical feedback and a kind of initial informal consensus that an idea is worth pursuing further. More commonly, contemporary research requires teams of researchers, with each team member contributing specific skills that may not be held by others in the team. Writing a research proposal for such projects necessarily involves coming to a consensus among all the scientists involved.

In addition to these informal consensus processes in formulating ideas and producing research proposals, there are other cases where more formal consensus methods are used to generate research ideas, to establish research priorities or to set out required directions for future progress in an area. Below are three case examples illustrating how formal consensus methods have been used to rank priorities (Case Examples 2.1 and 2.2) and to select the most promising interventions to test in an experiment (Case Example 2.3).

Case Example 2.1: British Psychological Society's Statement on Research Priorities for the COVID-19 Pandemic

The COVID-19 pandemic, which began in 2019, required a rapid scientific response to reduce its health and social impacts. In 2020, the British Psychological Society convened a group of nine experts to develop research priorities for psychological science in relation to the pandemic (O'Connor et al., 2020). These experts represented a range of areas within the discipline and were assisted by a wider advisory group of 19 psychological scientists.

(continued)

Case Example 2.1: (continued)

The priorities were generated through a series of ten long face-to-face meetings of the core group and discussions with the wider advisory group. An online survey was also carried out with a larger group of 539 psychological scientists to check whether the core and advisory groups had missed any key research priorities and to identify the highest-ranked priorities of those identified by the core group. This consensus process identified 18 priorities which were grouped into 7 domains:

1. Groups, cohesion and conflict
2. Working environment and working arrangements
3. Children and families
4. Educational practices
5. Mental health
6. Physical health and the brain
7. Behaviour change and adherence

The expert group made a "call to action" for psychological scientists to work collaboratively to research the identified topics.

Case Example 2.2: Determining the World Health Organization's Priorities for Research and Development on Emerging Infectious Diseases

New infectious diseases periodically emerge and pose a threat to health worldwide. WHO has developed a methodology, called the *R & D Blueprint*, which aims to prioritize efforts to make medical technologies available for emerging diseases for which few or no countermeasures exist. To implement this in 2017, WHO selected a group of 24 experts with diverse areas of expertise and geographic coverage (Mehand et al., 2018). The group developed a short-list of 13 potential priority diseases. The experts then independently rated each of the diseases for priority. Because some of the diseases were insufficiently differentiated, a second round of rating was carried out. The result was a list of six top-ranking diseases: Ebola virus infection, Marburg virus infection, Middle East Respiratory Syndrome coronavirus infection, severe acute respiratory syndrome, Zika virus infection and Crimean-Congo hemorrhagic fever. WHO has used these priorities to guide research and development on dealing with these diseases.

Case Example 2.3: Selecting Interventions for the Strengthening Democracy Challenge

The Strengthening Democracy Challenge is a project set up by American social scientists in response to the increasing polarization in US politics. It aims to reduce the American public's animosity towards political opponents. A very large experiment was carried out to test interventions that aimed to reduce partisan animosity, support for undemocratic practices and support for partisan violence. To identify potential interventions that could be tested in the experiment, the researchers made an open call for suggestions from social scientists (Voelkel et al., 2023). The interventions had to be easy to implement, brief, inexpensive and scalable (e.g. short online videos). The researchers received 252 submissions on potential interventions, but only had funding to test 25. To select the most promising 25, a multistage consensus process was carried out by the researchers. Each of the interventions was rated on a five-point scale by a subset of the team, which was used to reduce the list down to 70 interventions. Each of these 70 was reviewed by a further two evaluators to reduce the list to 50. Then a team of seven evaluated all 50 to reduce the list to the final 25. Once consensus processes agreed on the 25 interventions, these were tested in a megaexperiment with over 32,000 participants. Partisan animosity was most reduced by highlighting sympathetic and relatable individuals with different political views, while support for undemocratic practices and partisan violence was most reduced by correcting misperceptions about the views of political opponents.

Use of Consensus to Assess Funding Applications and Access to Resources

Writing an application for funding or other resources requires sign-off by all members of the research team and necessarily reflects a consensus. Once an application is written, it must be assessed by an agency that provides the resources.

The number of research proposals that scientists generate generally far exceeds the financial and other resources available to implement them. For this reason, a prioritization process is needed to select the higher quality ones. These processes usually involve expert consensus in some form. In order to select the best quality proposals, some sort of criteria need to be set for what constitutes "quality" and procedures are needed for evaluating proposals according to these criteria. Funding and other resource allocation agencies use committees of experts to develop these criteria and the procedures for implementing them. The actual selection of the highest quality proposals is also typically carried out by committees of experts using the criteria. Below are two examples, one illustrating consensus processes in selection of proposals for funding (Case Example 2.4) and the other concerning allocation of time using a piece of major equipment (Case Example 2.5).

Case Example 2.4: The Australian Research Council's Processes for Allocating Research Funding

Government funds available for research are generally insufficient to cover all applications for funding. Therefore, processes are needed to rank applications for priority. The Australian Research Council (ARC) is an Australian government agency that funds fundamental and applied research in all disciplines other than medicine. The Council runs a number of different funding schemes, all of which involve rankings by expert committees (Australian Research Council, 2024). The ARC advertises grants available and publishes guidelines for applicants. Applications are submitted by individual researchers or teams and receive assessments by experts. Assessors use a scoring matrix to score applications against the selection criteria. The applicants are given the opportunity to respond in writing to the assessors' comments. The applications are then assessed by a Selection Advisory Committee, which may have disciplinary sub-panels. The Committee ranks applications and recommends budgets for the highly ranked applications. The CEO of the organization then considers whether the applications are in the national interest and passes the final recommendations to the Minister for Education for approval.

Case Example 2.5: Allocation of Time for Using the Hubble Space Telescope

Research in astronomy requires access to expensive telescopes and the time available for their use is not sufficient to meet the demand from astronomers. An example is the time available to use the Hubble Space Telescope (Chawla, 2021). In 2014, the amount of time requested was six times the time awarded. To get time using the telescope, astronomers have to submit a proposal to the Space Telescope Science Institute, which manages the telescope, and these proposals are evaluated by the Institute's Time Allocation Committee. The Committee is subdivided into panels that review proposals within specific astronomical categories, such as stellar populations, solar system objects and cosmology. To be successful, a proposal has to have high scientific merit and also demonstrate that the observations required are only possible with the Hubble Telescope's unique capabilities. Proposals are also assessed according to their time requirements, with projects requiring shorter times being able to be slotted into gaps between those requiring longer observations. Since 2018, the Institute has used a double-blind system for evaluating proposals, in which both the applicants and the reviewers are blinded to each other's identities. This blinding was introduced to reduce any gender or other biases in evaluations. The Committee votes on the proposals and provides a recommended list to the Institute Director for final approval.

Use of Consensus When Implementing a Research Project

In carrying out a research project, scientists need to use methods that are accepted by other scientists. For example, when a scientific concept is used, there needs to be an accepted interpretation of its meaning so that scientists can communicate clearly with each other. Similarly, when a measurement is made, this needs to be equivalent to measurements from other laboratories to allow comparison between researchers and pooling of data.

Researchers also need to agree on the standards for a high-quality implementation of a research design, covering issues like how to frame hypotheses, how to use control conditions, and how to use blinding when taking measurements. There are a range of expert consensus statements to guide how various research designs should be implemented, which are widely adhered to. If scientists adhere to these standards, they know that other scientists are more likely to accept their findings and conclusions.

Below are a number of case examples, drawn from a range of scientific disciplines, illustrating the use of consensus methods to standardize measures (Case Examples 2.6, 2.7 and 2.8), to standardize terminology and define concepts (Case Examples 2.8 and 2.9) and describe the requirements for a methodologically rigorous research protocol (Case Example 2.10).

Case Example 2.6: A Protocol for Measuring the Volume of the Hippocampus with Magnetic Resonance Scans

For scientists to be able to compare the findings from different research projects, they need to ensure that the methods used for measuring variables are standardized. One area where standardization has been a problem is in the measurement of brain structures. The hippocampus is a brain structure that plays an important role in memory and atrophies in Alzheimer's disease. The volume of the hippocampus can be measured using MRI scans. However, there are various protocols for defining the boundaries of the hippocampus, leading to inconsistencies in volume measurements between researchers. To overcome this problem, The European Alzheimer's Disease Consortium (EADC) and the Alzheimer's Disease Neuroimaging Initiative (ADNI) cooperated to develop a 'Harmonized Protocol' for defining the boundaries of the hippocampus on MRI scans (Boccardi et al., 2015). The developers gathered 12 existing protocols and analysed them for which anatomical sub-structures were included in the boundaries of the hippocampus. Seventeen experts on the anatomy of hippocampus were recruited and voted across several voting rounds on which sub-structures should be included. After each voting round, the experts were asked to give reasons for their choices, and these were fed back to the other experts before the next voting round. The experts were also provided with data on how well each definition of the boundaries distinguished between Alzheimer's patients, people with mild cognitive impairment and normal controls. After five voting rounds, a consensus was reached. The new Harmonized Protocol was tested and found to produce very high agreement on volumes across different MRI raters and also very high reliability of the same raters across occasions.

Case Example 2.7: International Organization for Standardization (ISO) Standards for Physical Measurements and Laboratory Equipment

Comparison of scientific findings across studies and countries requires international standardization of physical measurements and laboratory equipment. The ISO is an international non-government organization representing over 160 national standards bodies (International Organization for Standardization, 2019). It brings together experts to develop consensus-based international standards. The ISO currently has over 22,000 standards, with approximately 100 new standards released each month. The standards cover a very wide range, including standards for credit cards, food safety practices, paper sizes, and codes for countries and currencies. Some of the standards are important to science, including standards for physical measurements (including volume, mass, density, viscosity, time, force, weight, fluid flow, acoustic measurements, electrical and magnetic measurements) and standardization of laboratory equipment (including laboratory devices, laboratory supplies, laboratory furniture and safety apparatus). The ISO standards are developed by groups of experts organized into technical committees. The experts are recommended by the ISO's national members. To develop a new standard, a proposal is put to a technical committee and voted on. If the proposal is approved, a working group is formed by the committee to prepare a draft. The draft is then voted on by the technical committee. If approved, the draft is circulated to all ISO members for comment and voting, with over two-thirds in favour needed to approve it. If there are no technical changes required, the standard is approved.

Case Example 2.8: Working Groups of the International Astronomical Union on Nomenclature and Measurement

Communication among scientists requires that they use technical terms in an agreed way. This is seen in astronomy, where names and categories of astronomical bodies need to be standardized. The International Astronomical Union (IAU) is an organization of professional astronomers from across the world which has a role in such standardization. The IAU establishes working groups with a minimum of five members to undertake certain tasks for limited time periods (International Astronomical Union, 2023). Some of these working groups deal with standardization of nomenclature and measurement. For example, there are working groups that publish recommendations on Planetary System Nomenclature, Exoplanetary System Nomenclature, Star

(continued)

Case Example 2.8: (continued)

Names, Time Metrology Standards, and Cartographic Coordinates and Rotational Elements. An example of the IAU's work was the 2006 reclassification of Pluto as a dwarf planet rather than a planet. This decision was based on a new definition of a planet drafted by committee and approved by the IAU General Assembly (International Astronomical Union, 2006). The definition is: "A planet is a celestial body that (a) is in orbit around the Sun, (b) has sufficient mass for its self-gravity to overcome rigid body forces so that it assumes a hydrostatic equilibrium (nearly round) shape, and (c) has cleared the neighbourhood around its orbit" (p. 4). Pluto failed to meet the third of these criteria.

Case Example 2.9: The International Classification of Diseases

Agreed definitions of diseases are needed to allow comparison of mortality and morbidity data across health practitioners, researchers, health facilities and countries. The International Classification of Diseases (ICD) has been developed by the World Health Organization (WHO) to define disease categories. The classification is used for coding of official government statistics on mortality and morbidity, for defining diseases in research studies and by clinicians when diagnosing diseases in patients. By having an international system of classification, it is possible to compare data on diseases from all countries across the world. The ICD is revised periodically by WHO, with its 11th edition (ICD-11) being implemented in 2022 (Pezzella, 2022). The ICD-11 consists of 26 chapters corresponding to groups of diseases. The WHO assigned the responsibility for developing the chapters to International Advisory Groups. For example, in order to revise the chapter on mental and behavioural disorders from ICD-10 to ICD-11, the WHO appointed a group of international experts to oversee the task. The advisory group in turn established working groups with relevant scientific and clinical expertise for each category of mental disorder, for example anxiety disorders and mood disorders. The task of the working groups was to review the relevant research evidence and recommend specific changes to their ICD section. The working groups tested out any draft changes by carrying out field studies in various clinics across the world. The working groups then modified their drafts based on feedback from clinicians. The ICD-11 was adopted by consensus of the World Health Assembly in 2019 and came into effect in 2022.

Case Example 2.10: SPIRIT Statement on the Minimum Content of a Clinical Trial Protocol

When researchers use a particular research method, they need to ensure that they are conforming to best practice in implementing that method. Clinical trials are widely used method in medical research for investigating the efficacy of treatments. An early step in carrying out a clinical trial is to write a trial protocol. To facilitate the writing of good-quality protocols, the Standard Protocol Items: Recommendations for Interventional Trials (SPIRIT) 2013 Statement was developed by a 19-member SPIRIT group to specify the minimum content that should be included in a protocol (Chan et al., 2013). The statement provides a 33-item checklist for items that should be included, such as study setting, trial design, interventions used, participant eligibility criteria, sample size, outcomes measured, data collection methods and statistical methods. Development of the checklist started with a systematic review of existing protocol guidelines, which was used to produce a preliminary checklist of 59 items. The preliminary checklist was refined using an expert consensus method in which 96 expert panelists rated each item for importance on a 1 to 10 scale. Panelists could provide comments to support their ratings, which were fed back to other panelists. They could also suggest additional items which were included in subsequent surveys. Over three survey rounds, panelists came to a consensus about the highest priority items, defined as those with a median score of 8 or higher. A second systematic review was carried out on each of the protocol items to see whether they contributed to trial quality, and the checklist was piloted with graduate students who were writing trial protocols. The final version of the statement was produced by consensus of the SPIRIT group using this information.

Apart from the use of expert consensus in carrying out scientific research, consensus also plays a role in matters of values, such as ethics. Before any project involving human or animal participants is undertaken, it has to be approved by an ethics committee, which uses an expert consensus approach. Ethics committees are themselves guided in their deliberations by published statements on ethical principles in research and procedures for reviewing projects, and these statements are themselves the product of a committee consensus. Committees may also produce statements on good practice in science and dealing with scientific misconduct. Case Example 2.11 describes the way consensus was used to develop a national statement on responsible conduct of research.

Case Example 2.11: The Australian Code for the Responsible Conduct of Research

Agencies that fund research and institutions where research is carried out have an interest in ensuring that scientists act to high standards of integrity. In Australia, the nation's two major research funding organizations and its universities have developed the 2018 Australian Code for the Responsible Conduct of Research (Kelso, 2016). The Code has guidance for researchers on responsible research practices in areas like authorship, management of data, peer review, disclose of interests and management of conflicts of interest, supervision, collaborative research, and publication and dissemination of research. It also provides guidance on managing and investigating potential breaches of the Code. To develop the Code, a draft was written by a committee of nine individuals, made up of experts from a range of disciplines, as well as experts on ethics and research integrity. Around 100 organizations with an interest in research were invited to provide feedback on the draft and 48 submissions were received. This feedback was used by the expert committee to write a final version of the Code.

2.2 Use of Consensus in Publishing Findings

When findings are published, there are generally standards on how these should be reported. Journals have guides for authors, which generally cover types of articles, standards for reporting different types of methodology, required section headings, use of language, formatting of tables and figures, referencing and acknowledgements of contributions to the research. Expert consensus groups may be used to produce these. An example in the area of medical research is the recommendations on publication by the International Committee of Medical Journal Editors (Case Example 2.12).

In addition to these general reporting guidelines, there are many standards for reporting specific methodologies or technologies in scientific publications, which have been developed through expert consensus. Two examples are the CONSORT Statements on standards for reporting clinical trials (Case Example 2.13) and consensus recommendations on reporting data from Magnetic Resonance Spectroscopy (Case Example 2.14).

Case Example 2.12: Recommendations of the International Committee of Medical Journal Editors (ICMJE)

Editors of scientific journals are important gatekeepers of what research is published. They must ensure that submitted papers are properly peer reviewed and that they meet acceptable standards of conduct and reporting. In some disciplines, editors have cooperated to develop agreed-upon standards. The ICMJE is a working group of editors of 16 general medical journals that meets annually. The ICMJE publishes recommendations for the conduct, reporting, editing and publication of scholarly work in medical journals. ICMJE (International Committee of Medical Journal Editors, 2023) developed these recommendations "to review best practice and ethical standards in the conduct and reporting of research and other material published in medical journals, and to help authors, editors, and others involved in peer review and biomedical publishing create and distribute accurate, clear, reproducible, unbiased medical journal articles" (p. 1). The ICMJE recommendations are widely followed by medical journals, beyond those of the 16 journals involved.

Case Example 2.13: Consolidated Standards of Reporting Trials (CONSORT) Statements

The CONSORT Statement was originally developed to give guidance on the minimal standards for reporting the results of clinical trials. However, there are now other CONSORT Statements on the reporting of a range of methodologies used in medical research and related disciplines, including health economic studies, case studies, qualitative studies and animal pre-clinical studies. All of these were developed using expert consensus. An example is the CONSORT Statement on the reporting of clinical trials on interventions involving artificial intelligence (Liu et al., 2020). The development of this statement involved a survey of stakeholders, a two-day consensus meeting of stakeholders and piloting of a draft checklist with multiple participants.

Case Example 2.14: Minimum Reporting Standards for In Vivo Magnetic Resonance Spectroscopy

Magnetic resonance spectroscopy (MRS) involves the use of an MRI scanner to study biochemical processes in the body. Researchers were concerned that when MRS research was reported, there was often a lack of detail about the methods used, which made it difficult for other researchers to critically evaluate the results, to replicate findings and to compare findings from different studies. For this reason, a consensus group of MRS experts produced minimum standards for reporting MRS results, including the hardware used, how data were acquired, methods of analysis and quality assessment (Lin et al., 2021). To develop the standards, an initial group of 21 MRS experts drafted, discussed and edited the standards document. A wider group of 19 experts was then recruited to a working group to support the recommendations.

Expert consensus also plays a role in publication through peer review of submissions to scientific journals. Peer review acts as a filter to ensure that only high-quality research is published and helps authors to improve the quality of manuscripts that are publishable. When a manuscript is submitted for publication, an editor will do an initial check that it is suitable to the focus of the journal and that it meets minimal standards before inviting a number of experts in the field to review the manuscript. Reviewers will be asked to comment on the originality of the research, the quality of the methodology, and the standard of the writing. They will also typically be asked to make an overall recommendation on whether or not the manuscript is publishable. The editor may decide to accept the manuscript as it is or, more typically, to invite the authors to revise the manuscript to take account of the reviewers' comments. For a manuscript to be accepted for publication would generally require a consensus from all reviewers and the editor. The limitations of peer review have been frequently noted, including the low inter-rater reliability of assessments (Bornmann et al., 2010), but it remains almost universal across many areas of science.

Journal editors play a key role in implementing reporting standards, organizing peer review, and making decisions about which submissions to publish. This role requires specific knowledge and skills in order to be done well. As shown in Case Example 2.15, expert consensus has been used to determine the core competencies of editors of biomedical journals, which can be used to guide the training of scientists moving into editorial roles.

Case Example 2.15: Core Competencies for Scientific Editors of Biomedical Journals

Editors of scientific journals play an important role in maintaining the standards of published research. This is a highly skilled role carried out by very experienced scientists, but the competencies required are not well defined. To fill this gap, a group of experts on scientific editing and publishing came together to develop the core competencies for scientific editors of biomedical journals (Moher et al., 2017). They began by carrying out a scoping review of published literature on the topic and extracted a list of potential competencies from this literature. They then surveyed journal editors about their perceptions of competencies required and training needs. These steps resulted in a potential list of 230 competencies. A Delphi expert consensus study was undertaken with 105 editors to find out which of the 230 potential competencies were most highly rated, which reduced the list down to 109. A series of consensus meetings was then held with stakeholders, who reduced the list down to a final list of 14 core competencies. These core competencies cover the qualities and skills of people who are selected to become editors, competencies needed on publication ethics and research integrity, and competencies on editorial principles and processes. Examples of the competencies are that scientific editors are able to: "Identify situations in which knowledge or skill required exceeds their level of competency and seek help or advice from appropriate colleagues or organizations"; "Describe what constitutes a breach in publication ethics, action on allegations of misconduct, misbehavior, or questionable practices, and proceed to issue an erratum or retraction when it is warranted, maintaining confidentiality, fairness, and due process"; and "Identify the vision and mission (aim and scope) of their journal and determine whether submitted manuscripts align with them" (Moher et al., 2017, pp. 5–7). These competencies have been endorsed by a range of scientific publishing organizations.

2.3   Use of Consensus When Reviewing the Literature

Literature reviews are used to critically evaluate and integrate the findings across studies on a topic. Some reviews give a selective coverage of what the reviewer believes are the notable findings, whereas others take a systematic approach and aim to provide an unbiased coverage of the whole literature on a topic. Some systematic reviews use meta-analysis to statistically pool data from multiple studies to give a quantitative estimate of effects. A number of expert consensus standards have also been developed for reviewing the literature on primary studies and these are widely used in some disciplines, particularly health and social sciences. Case Example 2.16 involves development of the PRISMA statements, which are used to guide how systematic reviews are reported, while Case Example 2.17 describes the development of AMSTAR, which is used to evaluate the quality of systematic reviews.

Case Example 2.16: PRISMA Statement on the Reporting of Systematic Reviews and Meta-analyses

Reviews of the scientific literature need to report the methods used for the review in sufficient detail so that they can be assessed and replicated by other scientists. The Preferred Reporting Items for Systematic Reviews and Meta-Analyses (PRISMA) Statement was introduced in 2009 to improve the reporting of systematic reviews and meta-analyses (Page et al., 2021). It guides authors on how to transparently report why a review was done, how it was done and what was found, by providing a checklist of 27 items that should be reported. An updated version, PRISMA 2020, was published in 2021. In order to update PRISMA, the developers reviewed existing guidelines for reporting systematic reviews and used this to inform the content of a survey on possible changes to the 27 PRISMA items. Systematic review methodologists and journal editors were invited to complete the survey online, with 110 completing it. The proposed modifications and the results of the survey were discussed at a two-day in-person meeting of 21 experts. An initial draft was produced and circulated to the other participants and experts for feedback over several revisions. The final version was approved by the 26 experts who were the co-authors of the revised Statement.

Case Example 2.17: Assessing the Quality of Systematic Reviews Using the AMSTAR Checklist

Systematic reviews of the scientific literature need to be carried out to a high standard so that other scientists can assess whether the conclusions can be relied upon. To help achieve this, the Assessment of Multiple Systematic Reviews (AMSTAR) checklist was developed as a way of measuring the quality of systematic reviews (Shea et al., 2007). The team responsible for developing AMSTAR made an initial list of 37 potential items by combining two existing checklists and adding some additional items that they thought were important. They then used this draft checklist to appraise 99 systematic literature reviews. The statistical method of factor analysis was used to reduce the 37 items down to 11 underlying dimensions of quality. The team then convened a meeting of 11 experts in the fields of methodological quality assessment and systematic reviews. The experts were shown the results of the factor analysis to inform their decision-making. They then discussed and independently voted on which items to include in the final checklist and what label should be given to each item. The 11 agreed items were fine-tuned by the group, leading to a final version of the checklist.

There are a number of organizations that specialize in carrying out systematic reviews to a very rigorous standard. The Cochrane Collaboration (Chandler & Hopewell, 2013) and JBI (Barker et al., 2023) do these for the medical sciences, and the Campbell Collaboration (Chandler et al., 2017) does the same for the social sciences. All three organizations use standardized methods that have been developed by expert groups and are incorporated in handbooks and tools used by reviewers. In Chap. 1, I noted that JBI has a scheme for rating the level of evidence for the effectiveness of health interventions, in which expert consensus receives the lowest rating. Paradoxically, JBI's level of evidence rating scheme was developed using the consensus of an expert working group. JDI is not alone in this regard. A number of other consensus-based frameworks for rating levels of evidence give a similarly low rating to consensus, even though consensus provides the foundation for their methodology.

2.4   Use of Consensus for Conclusions on Facts

Finally, consensus processes can be used to draw conclusions about scientific facts. In many cases, a consensus among scientists about the facts in their area will arise spontaneously as they become persuaded by key evidence. However, for more complex scientific questions, a formal consensus process may be carried out. Case Example 2.18 presents an example of a formal consensus process being used to agree on facts about low-calorie sweeteners, while Case Example 2.19 involves determining the consensus of scientific opinion on the accuracy of eyewitness memory in police and court proceedings. Another example is the consensus process of the Intergovernmental Panel on Climate Change to estimate the degree of global warming, which is discussed in more detail in later chapters.

Case Example 2.18: Facts on Low-Calorie Sweeteners

There has been debate about the safety and potential benefits of low-calorie sweeteners. Following a 2018 international conference on the topic, a consensus workshop was held in order to establish what is known (facts), what requires more research (gaps) and how progress might be achieved (actions) (Ashwell et al., 2020). The workshop involved 17 international experts who were speakers or chairs at the conference. The experts had diverse expertise in various aspects of low-calorie sweeteners. To identify facts, the experts rated statements prepared by the workshop convenors on a scale from 1 = strongly disagree to 10 = strongly agree. Statements that had a high level of endorsement were discussed further to come to a consensus on the wording. Following are the examples of facts agreed on:

- Intervention studies have shown that beverages containing low-calorie sweeteners have at least a similar effect on appetite and energy intake to water.
- The collective evidence supports the conclusion that there is no relationship between adiposity and liking/preference for sweet taste in either adults or children.
- The collective evidence supports the conclusion that low-calorie sweeteners have no adverse effect on blood glucose and insulin regulation in individuals with, and without, diabetes.

Case Example 2.19: Expert Opinion About the Science of Eyewitness Memory

Eyewitness testimony is frequently used by the police and courts. Surveys of experimental psychologists knowledgeable about eyewitness memory have been carried out over the past several decades to inform these stakeholders about the generally accepted scientific opinion in the field. The most recent survey recruited 76 scientists who had published peer-reviewed research on the topic or were members of relevant professional organizations (Seale-Carlisle et al., 2024). These scientists were asked to rate their agreement with 24 statements, with 14 of these receiving 80–100% agreement. For example, there was 100% agreement that "An eyewitness's testimony about an event often reflects not only what they actually saw but also any information they have learned since the event" and "When an eyewitness is questioned, how the question is worded can influence the eyewitness's answer". When the findings were compared to previous surveys, opinions were largely consistent over time, but there was greater nuance about some issues. For example, in the past witness confidence was considered to be a poor indicator of memory accuracy, but scientists now believe that it can be an indicator of accuracy under some circumstances.

2.5   The Multiple Uses of Expert Consensus in One Project

So far, this chapter has taken the reader through various stages of the scientific process, one step at a time, showing how consensus is often involved. To give a more holistic picture of how consensus can pervade a project, I will now go through a specific research programme that I have been involved in, pointing out the multiple uses of expert consensus. I have chosen this example because, having been personally involved in all phases of the project, I am acutely aware of how consensus was important at many points throughout.

The aim of this research was to improve the quality of parenting towards teenagers in order to reduce their risk of developing depression and anxiety disorders. It is known that adolescence is a period of life where the prevalence of depression and anxiety disorders rises sharply. It is also known that certain styles of parenting increase risk, while other styles reduce it. The research involved developing an online intervention for parents of teenagers, called *Partners in Parenting*, which was hypothesized to improve the quality of parenting and thereby prevent depression and anxiety problems in teenagers (Yap et al., 2017).

I have identified ten points in this research where expert consensus had some role in the project. Sometimes the research team directly used a consensus method, while at other points we relied on the products of consensus processes used by others. To indicate each point where consensus was involved, I have numbered them from #1 to #10.

To guide the development of the online intervention, we first carried out a systematic review and meta-analysis of studies examining parental factors associated with depression and anxiety disorders. This review used the PRISMA Statement to guide its methodology (#1; see Case Example 2.15). Synthesizing the findings from 181 studies, we found a higher risk in adolescents whose parents showed less warmth, more conflict between the parents, were over-involved with their child, had aversive interactions with their child, gave their child less autonomy and took less in interest in what their child was doing (Yap et al., 2014).

While this review told us about general factors that affected an adolescent's risk, it did not describe particular actions that parents should take to reduce risk. For example, if being a warm parent decreases risk, what specifically should a parent do to show this warmth? To fill out this detail, we carried out a Delphi expert consensus study (Yap et al., 2014). We did a literature search to find specific parenting recommendations and came up with a list of 402 of these. We constructed a survey questionnaire consisting of items on these parenting recommendations. The questionnaire included parenting actions such as "Parents should regularly show physical affection for their child, e.g. with hugs" and "Parents should discourage siblings from putting down or teasing each other". The questionnaire development required many meetings with all five investigators, who had to reach consensus on the wording of each item (#2). We recruited an international panel of 27 experts to rate each item for its preventive importance. The experts were also provided with the findings from the systematic literature review to guide their judgements. The experts came to a consensus (at least 90% agreeing that it was important) for 190 of these parenting strategies (#3).

The 190 endorsed strategies were then used to guide the content of the *Partners in Parenting* intervention. We developed an online questionnaire (the *Parenting to Reduce Adolescent Depression and Anxiety Scale*) that asks parents about their parenting behaviours in relation to the parenting strategies endorsed by the experts (Cardamone-Breen et al., 2017). The questionnaire data is used to give parents an automated feedback report on what they are doing well and on areas where they could improve. Based on areas recommended for improvement, the parents are offered a set of short online modules to build their skills in each specific area. The modules include interactive activities, real-life examples, audio clips, goal setting exercises and an end-of-module quiz. When constructing the modules, a reference group of parents was formed to give feedback on drafts (which is arguably another form of consensus, with potential consumers as experts).

The online intervention was then tested for efficacy in a randomized controlled trial, in which parents of adolescents were randomized to either Partners in Parenting or a control group where they received access to educational factsheets about adolescent development and mental health (Yap et al., 2019). To get funding for the trial, we successfully applied to a funding agency, Australian Rotary Health, which prioritizes applications using ratings from a committee of experts (#4). In writing the application, we used the SPIRIT Statement (#5; Case Example 2.10) to guide the details given about our methodology.

Participants in the trial were followed up over nine months. We found that the parents who received the online intervention reported better-quality parenting and that their child had fewer depressive symptoms. However, when the adolescents were asked about their own symptoms, no effect was found.

In writing up the findings from the trial, we followed the CONSORT Statement (#6; see Case Example 2.13). In matters of good research practice, such as decisions about authorship and data storage, we were guided by the Australian Code for the Responsible Conduct of Research (#7, see Case Example 2.11). Some of the journals we published the results in required us to follow the ICMJE guidelines (#8; see Case Example 2.12).

All our research was approved by a university research ethics committee, which operated by consensus (#9). The processes underlying this approval were guided by Australia's *National Statement on Ethical Conduct in Human Research*, which was developed by a consensus committee (#10).

This particular research project comes from a specific disciplinary area—psychological science. However, I believe it would be possible to do a similar analysis for projects in other disciplines. A number of the consensus elements seen in this project apply widely across many disciplines.

2.6   Summary of the Ways That Expert Consensus Is Used in Scientific Processes

The Case Examples presented in this chapter are a selection of many that could have been used. However, they illustrate some broader functions of expert consensus in scientific processes. I would group these as follows:

- Defining and standardizing concepts, terminology and measurements, for example naming astronomical bodies, defining disease categories and defining physical measurements.
- Agreeing on priorities, for example setting research priorities and allocating scarce resources.
- Setting methodological standards, for example creating hierarchies of evidence in medicine and setting standards for research designs such as clinical trials and systematic reviews.
- Judging quality of scientific outputs, for example peer review of grants and manuscripts, and checklists for judging quality of studies and reviews.
- Making value judgements, for example ethical standards and best practice codes.
- Agreeing on scientific facts.

Subsequent chapters elaborate in greater detail how consensus processes are essential. Chapters 3 and 4 examine the role of consensus in establishing scientific truth, while Chap. 5 looks at the role of consensus in translating research findings into practice and policy, including its role in making value judgements, and Chap. 6 examines its role in establishing what are acceptable scientific methodologies.

References

Ashwell, M., Gibson, S., Bellisle, F., Buttriss, J., Drewnowski, A., Fantino, M., et al. (2020). Expert consensus on low-calorie sweeteners: Facts, research gaps and suggested actions. *Nutrition Research Reviews, 33*(1), 145–154. https://doi.org/10.1017/S0954422419000283

Australian Research Council. (2024). *Overview of ARC funding process*. Retrieved August 14, 2024, from https://www.arc.gov.au/funding-research/peer-review/overview-arc-funding-process#assessment

Barker, T. H., Stone, J. C., Sears, K., Klugar, M., Leonardi-Bee, J., Tufanaru, C., et al. (2023). Revising the JBI quantitative critical appraisal tools to improve their applicability: An overview of methods and the development process. *JBI Evidence Synthesis, 21*(3), 478–493. https://doi.org/10.11124/JBIES-22-00125

Boccardi, M., Bocchetta, M., Apostolova, L. G., Barnes, J., Bartzokis, G., Corbetta, G., et al. (2015). Delphi definition of the EADC-ADNI harmonized protocol for hippocampal segmentation on magnetic resonance. *Alzheimer's & Dementia, 11*(2), 126–138. https://doi.org/10.1016/j.jalz.2014.02.009

Bornmann, L., Mutz, R., & Daniel, H. D. (2010). A reliability-generalization study of journal peer reviews: A multilevel meta-analysis of inter-rater reliability and its determinants. *PLoS One, 5*(12), e14331. https://doi.org/10.1371/journal.pone.0014331

Cardamone-Breen, M. C., Jorm, A. F., Lawrence, K. A., Mackinnon, A. J., & Yap, M. B. H. (2017). The Parenting to Reduce Adolescent Depression and Anxiety Scale: Assessing parental concordance with parenting guidelines for the prevention of adolescent depression and anxiety disorders. *PeerJ, 5*, e3825. https://doi.org/10.7717/peerj.3825

Chan, A. W., Tetzlaff, J. M., Altman, D. G., Laupacis, A., Gotzsche, P. C., Krleza-Jeric, K., et al. (2013). SPIRIT 2013 statement: Defining standard protocol items for clinical trials. *Annals of Internal Medicine, 158*(3), 200–207. https://doi.org/10.7326/0003-4819-158-3-201302050-00583

Chandler, J., Churchill, R., Higgins, J. P. T., Lasserson, T., & Tovey, D. (2017). *Methodological expectations of Campbell Collaboration intervention reviews (MECCIR): Reporting standards.* https://campbellcollaboration.org/meccir.html

Chandler, J., & Hopewell, S. (2013). Cochrane methods – Twenty years experience in developing systematic review methods. *Systematic Reviews, 2*, 76. https://doi.org/10.1186/2046-4053-2-76

Chawla, D. S. (2021, November 25). Record number of first-time observers get Hubble telescope time. *Nature.* https://doi.org/10.1038/d41586-021-03538-8

Crichton, M. (2003). Aliens cause global warming. In California Institute of Technology (Ed.), *Caltech Michelin Lecture.* Pasadena.

International Astronomical Union. (2006). *IAU 2006 General Assembly: Result of the IAU resolution votes.* IAU.

International Astronomical Union. (2023). *Working groups.* Retrieved February 22, 2023, from https://www.iau.org/science/scientific_bodies/working_groups/

International Committee of Medical Journal Editors. (2023). *Recommendations for the conduct, reporting, editing, and publication of scholarly work in medical journals.* ICMJE. www.icmje.org

International Organization for Standardization. (2019). In ISO (Ed.), *ISO in brief.* ISO.

Kelso, A. (2016). Review of the Australian code for the responsible conduct of research (2007). *Medical Journal of Australia, 205*(2), 49. https://doi.org/10.5694/mja16.00550

Lin, A., Andronesi, O., Bogner, W., Choi, I. Y., Coello, E., Cudalbu, C., et al. (2021). Minimum reporting etandards for in vivo magnetic resonance spectroscopy (MRSinMRS): Experts' consensus recommendations. *NMR in Biomedicine, 34*(5), e4484. https://doi.org/10.1002/nbm.4484

Liu, X., Cruz Rivera, S., Moher, D., Calvert, M. J., Denniston, A. K., Spirit, A. I., & Group, C.-A. W. (2020). Reporting guidelines for clinical trial reports for interventions involving artificial intelligence: The CONSORT-AI extension. *Lancet Digital Health, 2*(10), e537–e548. https://doi.org/10.1016/S2589-7500(20)30218-1

Mehand, M. S., Millett, P., Al-Shorbaji, F., Roth, C., Kieny, M. P., & Murgue, B. (2018). World Health Organization methodology to prioritize emerging infectious diseases in need of research and development. *Emerging Infectious Diseases, 24*(9). https://doi.org/10.3201/eid2409.171427

Moher, D., Galipeau, J., Alam, S., Barbour, V., Bartolomeos, K., Baskin, P., et al. (2017). Core competencies for scientific editors of biomedical journals: Consensus statement. *BMC Medicine, 15*(1), 167. https://doi.org/10.1186/s12916-017-0927-0

O'Connor, D. B., Aggleton, J. P., Chakrabarti, B., Cooper, C. L., Creswell, C., Dunsmuir, S., et al. (2020). Research priorities for the COVID-19 pandemic and beyond: A call to action for psychological science. *British Journal of Psychology, 111*(4), 603–629. https://doi.org/10.1111/bjop.12468

Page, M. J., McKenzie, J. E., Bossuyt, P. M., Boutron, I., Hoffmann, T. C., Mulrow, C. D., et al. (2021). The PRISMA 2020 statement: An updated guideline for reporting systematic reviews. *BMJ, 372*, n71. https://doi.org/10.1136/bmj.n71

Pezzella, P. (2022). The ICD-11 is now officially in effect. *World Psychiatry, 21*(2), 331–332. https://doi.org/10.1002/wps.20982

Seale-Carlisle, T. M., Quigley-McBride, A., Teitcher, J. E. F., Crozier, W. E., Dodson, C. S., & Garrett, B. L. (2024). New insights on expert opinion about eyewitness memory research. *Perspectives in Psychological Science,* 17456916241234837. https://doi.org/10.1177/17456916241234837

Shea, B. J., Grimshaw, J. M., Wells, G. A., Boers, M., Andersson, N., Hamel, C., et al. (2007). Development of AMSTAR: A measurement tool to assess the methodological quality of systematic reviews. *BMC Medical Research Methodology, 7*, 10. https://doi.org/10.1186/1471-2288-7-10

Voelkel, J., Stagnaro, M., Chu, J., Pink, S. L., Mernyk, J. S., Redekopp, C., et al. (2023). Megastudy identifying effective interventions to strengthen Americans' democratic attitudes. *Science, 386*(6719), eadh4764. https://doi.org/10.1126/science.adh4764

Yap, M. B. H., Cardamone-Breen, M. C., Rapee, R. M., Lawrence, K. A., Mackinnon, A. J., Mahtani, S., & Jorm, A. F. (2019). Medium-term effects of a tailored web-based parenting intervention to reduce adolescent risk of depression and anxiety: 12-month findings from a randomized controlled trial. *Journal of Medical Internet Research, 21*(8), e13628. https://doi.org/10.2196/13628

Yap, M. B., Lawrence, K. A., Rapee, R. M., Cardamone-Breen, M. C., Green, J., & Jorm, A. F. (2017). Partners in parenting: A multi-level web-based approach to support parents in prevention and early intervention for adolescent depression and anxiety. *JMIR Mental Health, 4*(4), e59. https://doi.org/10.2196/mental.8492

Yap, M. B., Pilkington, P. D., Ryan, S. M., & Jorm, A. F. (2014). Parental factors associated with depression and anxiety in young people: A systematic review and meta-analysis. *Journal of Affective Disorders, 156,* 8–23. https://doi.org/10.1016/j.jad.2013.11.007

Yap, M. B., Pilkington, P. D., Ryan, S. M., Kelly, C. M., & Jorm, A. F. (2014). Parenting strategies for reducing the risk of adolescent depression and anxiety disorders: A Delphi consensus study. *Journal of Affective Disorders, 156,* 67–75. https://doi.org/10.1016/j.jad.2013.11.017

3

Expert Consensus to Establish Scientific Truths

The previous chapter showed that consensus plays an important role in many scientific processes and hence refutes Crichton's (2003, p. 5) claim that "The work of science has nothing whatever to do with consensus". However, when laypeople and scientists criticize the use of expert consensus in science, their major objection is to its use to establish scientific truths. It would be possible to accept that consensus is involved in defining concepts and measures, agreeing on priorities, setting methodological standards and making value judgements, but not in determining what is truth. The alternative is that truth emerges directly from the findings of research projects and that we do not need consensus to know that the Moon causes the tides, that chlorophyl is essential to photosynthesis or that the heart pumps blood around the body.

In this chapter, I discuss in more detail whether expert consensus is an important indicator of scientific truth and perhaps even the best way available to establish what is scientific truth. However, I will concede that some of the critics do have a point and that scientists do not need to take formal votes or form working groups to decide on whether some scientific claims are true. However, that does not mean that consensus is not involved. In the next chapter, I will go on to argue that for some types of

© The Author(s) 2025
A. Jorm, *Expert Consensus in Science*, https://doi.org/10.1007/978-981-97-9222-1_3

scientific claims a consensus forms spontaneously, but that this may not be readily apparent to the outside observer. However, for other scientific claims, where the evidence is complex, formal consensus processes may be needed.

3.1   Positions Taken by Historians, Philosophers and Sociologists of Science

The role of consensus as an indicator of scientific truth has been discussed by a range of historians, philosophers and sociologists of science. For simplicity, I will refer to these as HPSS (history, philosophy and sociology of science) scholars. I will begin by taking an excursion into the range of positions they have taken. I do not attempt to review the positions taken by every scholar who has written something about the topic, but rather examine the positions of some major writers in the area showing the diversity of views. These positions can be arranged on a continuum, with HPSS scholars at one end of the continuum seeing consensus as playing a major role and those at the other end taking a negative position on its role.

3.1.1 The Positive Position

The most positive position on consensus for establishing scientific truth has been argued by Peter Vickers in his book *Identifying Future Proof Science* (Vickers, 2023). Vickers' interest is in identifying scientific facts that will be forever true (although perhaps subjected to minor fine tuning). His proposed criteria for identifying these are "a solid scientific consensus amounting to at least 95%, in a scientific community that is large, international, and diverse" (p. 18). The scientists must not merely agree with a scientific claim themselves but agree that it is an "established scientific fact", which is quite a high bar. Vickers proposes 30 examples of such lasting facts, such as "The Sun is a star", "Visual input coming from the retina is processed at the rear of the brain", "DNA has a double helix structure" and "At a constant temperature, the pressure of a gas is inversely proportional to its volume". Although he does not actually present data that at least 95% of the scientific community believe that these 30 examples are indeed established scientific fact, it seems plausible that they would have a very high level of endorsement. In the absence of such a survey, Vickers provides a general guide to whether the required level is reached:

> One good rule of thumb when trying to ascertain whether opinion has reached 95 per cent is this: in most cases where it has not, evidence of substantial debate in the community will be relatively easy to find, and in most cases where it has, any serious opposition (within the relevant scientific community) will be extremely difficult to find. (Vickers, 2023, p. 222)

The argument that Vickers makes for using expert consensus to ascertain lasting facts is that it is impossible for the non-specialist to look at the primary evidence, as there is far too much of it and it is difficult for anyone to evaluate unless they have very specialized expertise. On the other hand, it is much easier (but still difficult) to ascertain whether the relevant scientific community overwhelming accepts a given claim as true.

As a working scientist myself, I can relate to what Vickers says. I can evaluate the primary evidence in the disciplines in which I have done research, namely psychology and psychiatry, epidemiology and public health, gerontology, neurology and human genetics. However, for most scientific issues I do not have the time to read the primary evidence myself and have to rely on review articles written by others. For other disciplines beyond those I have worked in (e.g. economics and chemistry), I would generally be capable of only a basic understanding of the primary evidence, even if I had the time, and sometimes it would be well outside my competence. In these disciplines, I could spot pseudoscience and blatantly ridiculous work, but could not evaluate the detail of work that passes minimal standards. Even in areas where I do research, I may use methodologies that I do not fully understand the foundations of. I accept these methods because people with much greater expertise than mine have endorsed them. For example, I commonly use statistical methods without a knowledge of the mathematical proofs underlying them, or I have used techniques like MRI scanning of the brain without a full understanding of the underlying physics or physiology.

3.1.2 Conditionally Positive Positions

There are a number of HPSS scholars who see a role for consensus, but do not see it as being a strong indicator of truth, and certainly not as a definer of truth. However, for these scholars, there are certain conditions under which consensus may be formed which may make it a better indicator. These proposed conditions vary from author to author, but there are some consistencies. I will briefly summarize the views of four of these: Helen Longino, Naomi Oreskes, Aviezer Tucker and Boaz Miller. While these are not an exhaustive list of HPSS scholars who have argued for conditionally positive positions, they illustrate the range of views.

Helen Longino's (2002) book *The Fate of Knowledge* deals with a debate between two contrasting accounts of science, which she labels as "rational" and "social". She summarizes these two approaches as follows:

> Roughly, rational or cognitive approaches are those that focus on evidential or justifying reasons in accounting for scientific judgment. Social (or sociological) approaches, by contrast, focus either on the role of nonevidential (ideological, professional) considerations or on social interactions among the members of a community rather than on evidential reasons in accounting for scientific judgment. (p. 2)

Although these two approaches have often been seen as incompatible and competing, Longino argues that the division between the two is unnecessary because science has "a social character to its cognitive, or knowledge-productive, capacities" (Longino, 2002, p. 8). The observations and reasoning of individual scientists are supplemented by social processes involving criticism and survival of criticism involving other scientists. She writes:

> justification, or the production of knowledge, not just in the testing of hypotheses against data, but also in subjecting hypotheses, data, reasoning, and background assumptions to criticism from a variety of perspectives. Establishing what the data are, what the descriptive categories and their boundaries are, what counts as acceptable reasoning, which assumptions are legitimate and which not becomes a matter of social interactions as much as a matter of interaction with the material world. (Longino, 2002, p. 205)

For Longino, the social process of critical discussion is essential to science. Sometimes, scientists will come to a consensus, but this must be arrived at under certain conditions for it to be legitimate:

> Where consensus exists, it must be the result not just of the exercise of political or economic power, or of the exclusion of dissenting perspectives, but a result of critical dialogue in which all relevant perspectives are represented. (Longino, 2002, p. 131)

Historian Naomi Oreskes takes a similar position to Longino and argues that that scientific knowledge is fundamentally consensual. She is not as sanguine as Vickers about the possibility of establishing lasting scientific facts. In her book *Why Trust Science?* (Oreskes, 2019), she concludes that "the contributions of science cannot be viewed as permanent. The empirical evidence gleaned from the history of science shows that scientific truths are perishable" (p. 49). For her, consensus is more of a strong indicator rather than a definer of truth:

> we do not have independent, unmediated access to reality and therefore have no independent, unmediated means to judge the truth content of scientific claims. We can never be entirely positive. Expert consensus serves as a proxy. We cannot know if scientists have settled on the truth, but we can know if they have settled. In some cases where it is alleged in hindsight that scientists "got it wrong," we find on closer examination that there was, in fact, no consensus among scientists on the matter at hand. (Oreskes, 2019, p. 249)

Oreskes argues that the consensus of scientists is more likely to indicate the truth when certain conditions are met:

> objectivity is likely to be maximized when there are recognized and robust avenues for criticism, such as peer review, when the community is open, non- defensive, and responsive to criticism, and when the community is sufficiently diverse that a broad range of views can be developed, heard, and appropriately considered. (Oreskes, 2019, p. 53)

Unlike Vickers, Oreskes does not propose any criterion for defining consensus. However, her concerns are less about defining lasting truths, which arguably requires a very high threshold, than in providing scientific guidance on issues of social or health policy importance, like the role of human activity in climate change and the role of smoking in health, which she believes require a lower threshold.

Aviezer Tucker (2003) argues that a consensus of beliefs does not necessarily indicate knowledge and that knowledge can exist without consensus. What is important about any consensus is the factors that led to its development. Various hypotheses could be put forward to explain why a particular consensus developed. The important one to Tucker is the "knowledge hypothesis", which is that shared knowledge provides the best explanation of a consensus on beliefs. However, there are other potential hypotheses that might explain a consensus, for example shared biases, shared mistakes, political interests or coercion by some authority. The knowledge hypothesis and the alternative hypotheses need to be evaluated to see which provides the best explanation of the consensus:

> If all the alternative hypotheses to the knowledge hypotheses are false or are not as good in explaining a concrete consensus on beliefs, the knowledge hypothesis is the best explanation of the consensus. If the knowledge hypothesis is best, a consensus becomes a plausible, though fallible, indicator of knowledge. Though the knowledge hypothesis may be a better explanation of a consensus than all the competing hypotheses, this explanation is still fallible because a better hypothesis than the knowledge hypotheses is always possible. (Tucker, 2003, p. 504)

Tucker proposes three conditions under which a consensus may be formed, which make the knowledge hypothesis a more likely explanation for the consensus. These are: (1) the consensus is uncoerced (e.g. not due to intimidation, bowing to authority or economic dependence); (2) it is uniquely heterogeneous (e.g. the experts do not share an ideology, are from multiple cultures and gender diverse); and (3) it is sufficiently large, so as to make accidental results unlikely and to exclude hidden biases that may be in a small group.

Boaz Miller (2013, 2019) argues a position that is similar to Tucker's, namely that consensus is not necessarily an indicator of the existence of knowledge, but is more likely to be knowledge-based under certain conditions. He proposes the following three:

(1) social calibration: researchers give the same meaning to the same terms and share the same fundamental background assumptions; (2) apparent consilience of evidence: the consensus seems to be built on an array of evidence that is drawn from a variety of techniques and methods; and (3) social diversity: the consensus is shared by men and women, researchers from the private and public sectors, liberals and conservatives, etc. (Miller, 2019, p. 234)

Miller also discusses under what circumstances dissent from a consensus could be given less significance or dismissed:

dissent may be epistemically detrimental, especially dissent stemming from *manufactured uncertainty* or *doubt mongering*. Affluent bodies opposed to a particular piece of knowledge may inhibit the formation of consensus or create the perception that it does not exist. They may insist on more and ever more critical scrutiny, no matter how strong the evidence is. (Miller, 2019, p. 236)

Miller (2019) cites as an example climate change sceptics who are motivated by political or economic interests to promote dissent or create a perception that a consensus does not exist.

3.1.3 Negative Positions

Lastly, there are some HPSS scholars who take a largely negative position. I will outline the thinking of two of them: Miriam Solomon and Nicholas Rescher.

In her book *Social Empiricism*, Miriam Solomon (2001) argues that science commonly progresses without consensus and that consensus is unnecessary for scientific progress towards truth. This is because more than one theory can have truth in it. Consensus is desirable only in cases where one theory has all the available truths, which she claims is rare.

Solomon proposes that there are a set of factors, which she calls "decision vectors", that influence scientists' theory choices. Sometimes these factors will lead scientists towards consensus and other times towards dissent. Some of these factors she refers to as "empirical", because they make scientists prefer theories with empirical success. Others are "non-empirical", such as pride, peer pressure, acceptance of authority and biases in attitudes, which could push scientists either towards or away from theories with empirical success. While other HPSS scholars see the social interactions involved in achieving consensus as reducing the influence of such non-empirical factors, Solomon is sceptical that they do so. Her solution is difficult to follow but involves evaluating the range of factors that are influencing scientists' decisions ("decision vectors") in an area of scientific controversy and assessing the balance of "empirical" and "non-empirical" factors. Solomon's guidance on how this could be done and who should do it gives little practical advice for the working scientist, let alone the interested non-scientist, as the reader can judge from the following quote:

> Anyone who is situated so as to be able to both assess and influence the distribution of research effort—grant officers, science policy experts, sometimes department heads, journal editors—can do so as a social empiricist, in consultation with relevant experts on various decision vectors. This is a new locus of epistemic responsibility. Normative suggestions in philosophy of science are typically addressed to the individual working scientists. Social empiricism focuses, instead, on epistemic responsibilities at the level of policy. (Solomon, 2001, p. 150)

While Solomon's "empirical" and "non-empirical" factors have some resemblance to the conditions for reaching a better-quality consensus that other HPSS scholars have proposed, she gives no well-defined list of what these are that could be practically implemented.

At the most negative end of the continuum is Nicholas Rescher (1993) in his book *Pluralism: Against the Demand for Consensus*. His book is concerned with the use of consensus in society in general but does include consensus in science in its purview. Rescher's view is that pluralism of opinion is desirable and that we should not aim to achieve consensus on issues. Applied to the area of science, he states:

> Far from science being a domain pervaded by consensus, there is, in fact, good reason to think that dissensus and controversy are the lifeblood of scientific work at and near the frontiers of research—though, to be sure, the rational and social dynamics of scientific opinion formation does generally make for an eventual uniformization of scientific opinion. (Rescher, 1993, p. 40)

While the above quotation does acknowledge that scientific opinion will eventually move towards consensus, Rescher takes a negative view of how this occurs:

> What makes for consensus among the scientists of the day is not just (and perhaps not even primarily) the inherent rationality of 'the scientific method' seen as a bloodless abstraction of rational process. Rather, it lies in the operation of the social processes of the inquiring community. Scientists are impelled to consensus less by an intersubjectively rational methodology than by a conformism imposed by promotion committees, funding agency appraisers, and peer review boards. Yet even these pressures, though powerful, achieve only limited uniformity of thought in an area where innovation and novelty are of prime value. (Rescher, 1993, p. 42)

Rescher's concern is with areas of science where new ideas are emerging, rather than with the lasting scientific facts of Vickers. It would indeed be undesirable to have a forced or premature consensus in such areas. Nevertheless, Rescher does see at least some benefit in consensus processes, as a method for reducing errors and bias:

> For those consensual processes in matters of cognition are in the end no more than useful devices for eliminating or reducing mistakes of various sorts—mathematical calculating errors, for example, or experimenter bias. They are not so much mechanisms for assuring truth as safeguards against various particular sources of error. And since the elimination of error makes only a partial contribution to the discovery of truth, their operation leaves ample scope for disagreement and diversity. (Rescher, 1993, p. 38)

3.2   Is There Any Consensus Among HPSS Scholars?

It would be reassuring if HPSS scholars came to some agreement about the role of consensus in science, but the range of positions I have reviewed suggest that this may not be the case. It is, however, possible that HPSS scholars are actually in greater agreement than is apparent from the literature, as Vickers (2023) has suggested:

> There is a clear culture of critical thinking in philosophy, and it is not unusual for journal articles to emerge that do nothing more than criticise an idea somebody else has put forward, finding the gap in the argument, or the weak premise. And of course, one can't hope to publish a paper that merely agrees with an idea that has already been published. Thus when one looks to the literature one will see all sorts of disagreement, and perhaps little agreement, but this needn't mean that there aren't any interesting points of community agreement. It's just less obvious how exactly that agreement makes it into the literature. (p. 239)

Adding to the appearance of disagreement is that HPSS scholars are sometimes talking about different types of consensus. Vickers, for example, is interested in the use of consensus for establishing lasting scientific facts, whereas some of the other HPSS scholars are discussing consensus about much lower levels of belief (e.g. Longino, Tucker and Miller). There is a considerable difference between asking a scientist, "Do you think x is true?" and asking them, "Do you think x is a lasting scientific fact?" Taking my own areas of expertise as an example, if I were asked, "Do you think antidepressants work for severe depression?" I would say, "Yes", basing my response on a "balance of probabilities" assessment of the current evidence. However, if I were asked, "Do you think the efficacy of antidepressants for severe depression is a lasting scientific fact?" I would say, "No", because the bar of required evidence is much higher and there is some doubt about the matter. Those HPSS scholars at the negative pole on the role of consensus, who emphasize the importance of pluralism and dissent in science, appear to be considering scientific questions which are still highly contested. In such cases, any attempt to

Table 3.1 Illustration of how a very "High-Bar" scientific question endorsed at a very high level is a strong indicator of scientific truth

	Very "High-Bar" scientific question (e.g. Do you think x is a lasting scientific fact?)	"Low-Bar" scientific question (e.g. Do you think x is true?)
Very high level of endorsement	Strong indicator of scientific truth	Plausible claim
Lower level of endorsement (dissent)	Contested	Contested

produce a consensus on the truth would indeed be premature. Table 3.1 illustrates the possibilities. It shows a cross-tabulation of two dimensions: how high the bar is set in the question being evaluated and how high is the level of endorsement of that question. Where there is both a high-bar question and a very high level of endorsement, the consensus is arguably a strong indicator of scientific truth.

Despite the range of positions that HPSS scholars have taken about consensus, there are some areas of agreement about the conditions under which consensus is more likely to indicate truth. I see four recurring themes in these positions:

1. The consensus needs to be rational, empirical and critically examined.
2. The group coming to the consensus needs to be diverse.
3. The group needs to be open-minded, and there is no coercion of dissenters.
4. The group needs to be sufficiently large to get reliable results.

I have summarized these four themes in Table 3.2 and given illustrative supporting quotes from the various HPSS scholars I have reviewed. I will return to this again in Chap. 9, where I examine the evidence on the conditions needed for good-quality group judgements.

So, on the issue of whether consensus can be used to establish scientific truth, I would conclude that consensus can be a strong indicator of scientific truth under certain conditions. It depends on what question the experts are being asked, how the experts are chosen, and what processes they use to come to a consensus. If the question chosen is a very high-bar

Table 3.2 Some common themes in HPSS scholars' views about the conditions under which consensus is more likely to indicate truth

Theme	Supportive quotes from HPSS scholars
The consensus is rational, empirical, critically examined	"objectivity is likely to be maximized when there are recognized and robust avenues for criticism" (Oreskes, 2019, p. 53). "a result of critical dialogue in which all relevant perspectives are represented" (Longino, 2002, p. 131). "researchers give the same meaning to the same terms and share the same fundamental background assumptions... the consensus seems to be built on an array of evidence that is drawn from a variety of techniques and methods" (Miller, 2019, p. 234). "shared knowledge explains a consensus on beliefs" (Tucker, 2003, p. 504).
The group coming to the consensus is diverse	"a scientific community that is... international, and diverse" (Vickers, 2023, p. 18). "when the community is sufficiently diverse that a broad range of views can be developed, heard, and appropriately considered" (Oreskes, 2019, p. 53). "it must be the result not just of...the exclusion of dissenting perspectives" (Longino, 2002, p. 131). "social diversity: the consensus is shared by men and women, researchers from the private and public sectors, liberals and conservatives, etc." (Miller, 2019, p. 234). "The unique heterogeneity of a consensus group generates the strongest evidence against alternative hypotheses to the knowledge hypothesis" (Tucker, 2003, p. 506).
The group is open-minded; there is no coercion of dissenters	"when the community is open, non-defensive, and responsive to criticism" (Oreskes, 2019, p. 53). "it must be the result not just of the exercise of political or economic power" (Longino, 2002, p. 131). "An epistemically significant consensus must be uncoerced" (Tucker, 2003, p. 505).
The group is sufficiently large to get reliable results	"a scientific community that is large" (Vickers, 2023, p. 18). "The uncoerced heterogeneous group that reaches consensus must be sufficiently large to avoid accidental results" (Tucker, 2003, p. 512).

one, the experts are well chosen, the processes to determine consensus are ideal and the level of consensus approaches 100%, then I think that consensus can also establish scientific truth.

This chapter's excursion into the history, philosophy and sociology of science has dealt with the issue of what role, if any, consensus *should play* in determining scientific truths. In the next chapter, I turn to the issue of what role it actually *does play* in scientific practice, arguing that it is always present, but often the consensus processes are hidden from view.

References

Crichton, M. (2003). Aliens cause global warming. In California Institute of Technology (Ed.), *Caltech Michelin Lecture*. Pasadena.

Longino, H. E. (2002). *The fate of knowledge*. Princeton University Press.

Miller, B. (2013). When is consensus knowledge based? Distinguishing shared knowledge from mere agreement. *Synthese, 190*, 1293–1316. https://doi.org/10.1007/s11229-012-0225-5

Miller, B. (2019). The social epistemology of consensus and dissent. In M. Fricker, P. J. Graham, D. Henderson, & N. J. L. L. Pedersen (Eds.), *The routledge handbook of social epistemology* (pp. 230–239). Routledge.

Oreskes, N. (2019). *Why trust science?* Princeton University Press. https://doi.org/10.1515/9780691222370

Rescher, N. (1993). *Pluralism: Against the demand for consensus*. Clarendon Press; Oxford University Press. Publisher description http://www.loc.gov/catdir/enhancements/fy0639/93018392-d.html

Solomon, M. (2001). *Social empiricism*. MIT Press.

Tucker, A. (2003). The epistemic signficance of consensus. *Inquiry, 46*, 501–521. https://doi.org/10.1080/00201740310003388

Vickers, P. (2023). *Identifying future-proof science*. Oxford University Press.

4

Spontaneous and Deliberative Processes to Reach Consensus

Some of the critics of consensus in science quoted in Chap. 1 see truth emerging directly from the evidence without any need for a consensus process. On the surface, this view has some justification. Scientists have never had to resort to a consensus vote on the 30 established scientific facts that Vickers (2023) has proposed, yet these facts are widely accepted. However, I question whether the lack of a formal consensus process indicates that consensus does not play a role. I will argue that consensus does actually play a vital role in establishing such well-established facts, although this is often not apparent to an outside observer.

4.1 Spontaneous and Deliberative Processes

I propose that there are two contrasting processes by which consensus develops—which I call the "spontaneous" and "deliberative" processes. For ease of exposition, I discuss these as though they are quite separate processes, but they are better seen as ends of a continuum. Table 4.1 summarizes the characteristics of the two processes.

© The Author(s) 2025
A. Jorm, *Expert Consensus in Science*, https://doi.org/10.1007/978-981-97-9222-1_4

Table 4.1 Contrast between spontaneous and deliberative processes for developing consensus in science

Characteristic	Spontaneous consensus process	Deliberative consensus process
How consensus builds	Spontaneous adoption by scientists; relatively fast	Formal methods for determining consensus; relatively slow
Complexity of causality	Simpler causal processes; strong effects	Complex causal processes; weak effects
Complexity of the evidence base	Often from a single discipline; relatively easy for an expert to synthesize	Often spanning multiple disciplines and methodologies; requires many experts to synthesize
Size of the evidence base	Single or small number of very persuasive studies	Accumulation of evidence across many studies

The spontaneous process involves a consensus that develops rapidly and spontaneously among experts in an area. It will be manifested as a lack of substantial debate and serious opposition with the relevant scientific community (Vickers, 2023). Scientists working in the area will be aware of the consensus through the scientific literature and interactions with their peers. However, the process by which it builds may not be apparent to an outside observer who is not involved as a participating scientist and may lead the observer to think that the scientific facts emerge directly from the evidence without any need for consensus. An example of this is Curry and Webster's (2013) statement that "With genuinely well-established scientific theories, 'consensus' is not discussed and the concept of consensus is arguably irrelevant…there is no point in discussing a consensus that the Earth orbits the sun, or that the hydrogen molecule has less mass than the nitrogen molecule" (p. 3).

While it may not be explicit, there may be a number of indirect indicators, none of which are in themselves definitive. One of these is a rapid rise in citations. When scientists publish their findings, they cite previous publications that have informed the background to their research or which they wish to compare their findings to. Generally, these citations reflect a positive influence of the previous work, although sometimes they can involve criticism or rejection of earlier findings. A very high and rapidly rising rate of positive citations is an indirect indicator of acceptance by scientific peers and of a developing consensus. Other indicators of consensus are the award of prestigious prizes for important findings or theories, and incorporation into textbooks.

If citations are used as an indicator of consensus, it is important that these come from a diversity of other scientists. In small fields of research, it is possible for "citation cartels" to deliberately manipulate citations. This has been observed, for example, in the area of mathematics where groups of mathematicians in China, Saudi Arabia and other countries have repeatedly cited low-quality papers from colleagues at their own institution in order to improve their universities' rankings (Catanzaro, 2024). In such cases, a high rate of citations reflects a planned manipulation of data rather than a spontaneous consensus.

The deliberative process, by contrast, involves formal methods to develop the consensus, such as consensus conferences, expert working groups set up by international scientific organizations, Delphi consensus studies of expert opinion, and formal votes by groups of acknowledged experts. Given that the evidence is complex and involves many publications, there may not be any single one that is distinguished by a rapidly rising and very high rate of citations. High citations are more likely to be to consensus statements and systematic reviews of the literature rather than to the primary evidence.

The scientific questions that are subject to spontaneous consensus are more likely to involve simpler causality and strong associations between variables. By contrast, for questions subject to deliberative consensus, the causality will be complex, involving many variables. Because effect sizes tend to be smaller and subject to variability in estimation from study to study, systematic reviewing techniques and pooling of data through meta-analysis may be necessary to get greater precision.

For simpler causal processes and strong effects, it is often possible to answer a scientific question with a small number of critical studies showing replicable effects. In such cases, an individual scientist with relevant expertise would be able to read and critically evaluate the available literature and come to their own conclusion. Once the relevant scientific community has been able to evaluate the evidence, an implicit consensus may spontaneously occur. On the other hand, scientific questions that involve complex causality often involve an extensive literature of primary evidence, which may involve several disciplines and multiple methodologies. A consequence of this complex evidence base is that it is not feasible for an individual scientist to read it all, and certainly not with critical understanding. Therefore, to integrate and evaluate the evidence requires groups of cooperating scientists with diversity of expertise.

4.2   Illustrative Case Examples of Spontaneous and Deliberative Consensus

While there are many differences between the spontaneous and deliberative processes, the key indicator differentiating the two is how consensus is assessed, whether by a rapid rise in positive citations to key studies for spontaneous consensus or by a formal consensus process for deliberative consensus. To illustrate the continuum from spontaneous to deliberative consensus, I present three case examples below. Case Example 4.1 is from astronomy and illustrates a purely spontaneous process, with strong replicable effects in key studies receiving a rapid rise in citations and acknowledgement with a Nobel Prize within a short time.

Case Example 4.1: Accelerating Expansion of the Universe

Edwin Hubble's observations of galaxies in the 1920s showed that typically galaxies are receding from Earth, implying that the universe is expanding rather than static, and supporting the Big Bang theory. The conclusion that the universe is expanding raised the question of what its ultimate fate would be. Would gravity ultimately slow and then reverse the expansion so that there would be an eventual Big Crunch, or would it expand forever? Answering this question required the measurement of both the distance and the velocity of objects located at a range of distances from the Earth. Such objects had to have known intensities, so that the observed brightness could be used to infer the distances. Type Ia Supernovae have these properties, but are rare and transient, making the required observations difficult to achieve. However, computers could be programmed to detect their appearance by subtracting images of the same regions of space taken at different times. In the late 1990s, two teams of astrophysicists made these

---

(continued)

Case Example 4.1: (continued)

measurements: the Supernova Cosmology Project led by Saul Perlmutter and the High-Z Supernova Search Team led by Brian Schmidt and Adam Riess. To their surprise, these teams found that the expansion of the universe was actually accelerating.The first publication of these findings was made by Riess et al. (1998) followed by Perlmutter et al. (1999). There was no assessment of the scientific consensus on an accelerating expansion, but Kirshner (2013) noted that "Once the published results of the two teams concurred, a very rapid theoretical consensus precipitated, one that embraced this astonishing result of an accelerating universe" (p. 451), while also cautioning that "Of course, consensus of expert opinion does not imply that this is the correct view. That knowledge comes from more data" (p. 451). The rapid acceptance of an accelerating universe is shown by citation data, which was high in the year following publication and rapidly increased over the following years (see Fig. 4.1). In 2011, the finding was acknowledged by the awarding of a Nobel Prize to Perlmutter, Schmidt and Riess. The finding of an accelerating expansion matches the spontaneous consensus template, being driven by clear replicated results involving strong measurements and mathematical theory, with no formal assessment of scientific consensus needed. It took a little over a decade from its first publication to a Nobel Prize.

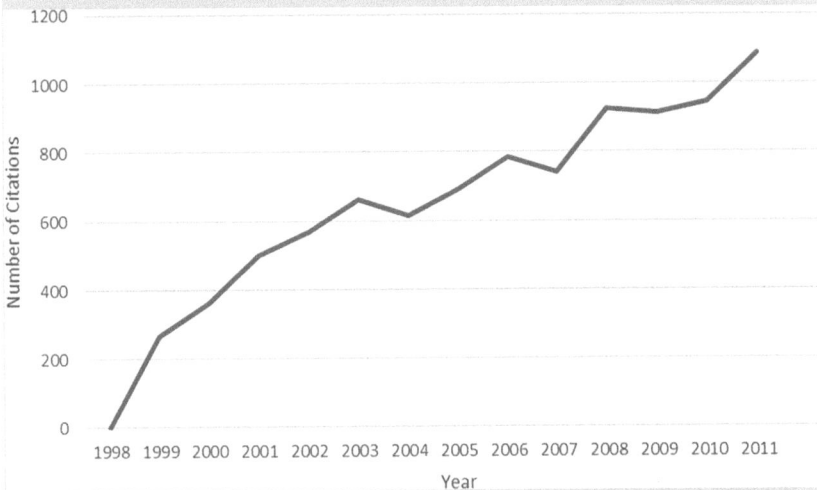

Fig. 4.1 Citations per year in Google Scholar to Riess et al. (1998). *Note.* Citations shown from year of publication (1998) to year of award of Nobel Prize (2011)

Case Example 4.2 comes from medical science and involves a largely spontaneous process as indicated by a rapid rise in positive citations leading to a Nobel Prize after two decades. However, there were some deliberative processes involved in translating the findings into clinical practice.

Case Example 4.2: Bacterial Infection and Stomach Ulcers

This case example involves the discovery that most stomach ulcers are caused by the bacterium *Helicobacter pylori* (Pincock, 2005). In the early 1980s, Barry Marshall and Robin Warren observed the presence of bacteria in the stomachs of patients with gastritis and thought these might have a causal role. Their initial observations were contrary to orthodox medical teaching of the time, which held that bacteria could not grow in the acidic environment of the stomach and that psychological stress played a major role. As a result, Marshall and Warren met with considerable scepticism from gastroenterology specialists. Their key paper, published in 1984, showed that the bacteria were present in almost all patients with active chronic gastritis, duodenal ulcer or gastric ulcer, indicating that they were likely to be causal. In the same year, Marshall deliberately infected himself with the bacterium and experienced a rapid onset of symptoms. Despite the initial opposition, there was rapid acceptance. Figure 4.2 shows the rapid rise in citations to the 1984 article. A decade later, in 1994, the US National Institutes of Health published a report on a consensus conference supporting the use of antimicrobial agents in ulcer patients with *H pylori* infection (National Institutes of Health, 1994), with other countries subsequently producing similar consensus statements (Lee & O'Morain, 1997). In 2005, Marshall and Warren received a Nobel Prize for their research. The consensus that bacterial infection causes stomach ulcers was largely spontaneous (a clear and strong causal link with a rapid acceptance indicated by positive citations), but has some elements of a deliberative process (e.g. formal consensus statements on clinical practice). Because Marshall and Warren's findings led to a new consensus about the cause and treatment of stomach ulcers, this has sometimes been given as an example how a consensus can be wrong (Briant, 2005). However, there continues to be empirical support for an additional role of psychological stress in the aetiology of stomach ulcers (Kanno et al., 2013; Levenstein et al., 2015). A consensus may therefore change without the previous consensus being completely false.

(*continued*)

Case Example 4.2: (continued)

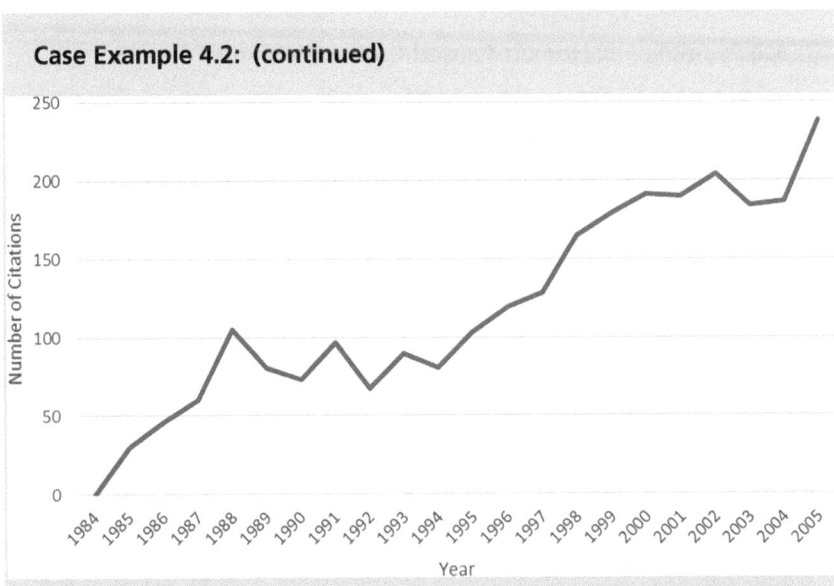

Fig. 4.2 Citations per year in Google Scholar to Marshall and Warren (1984). *Note.* Citations shown from year of publication (1984) to year of award of Nobel Prize (2005)

The third example, Case Example 4.3, involves the work of the Intergovernmental Panel on Climate Change (IPCC) to develop a consensus on the role of human activity in climate change. It is a clear example of a deliberate process, with a formal method for determining consensus, complex causality, and the involvement of many disciplines and a range of methodologies. The evidence base is too extensive and varied for any one scientist to critically evaluate. An international cooperative consensus process was needed and has led to progressively firmer conclusions over three decades.

Although both spontaneous and deliberative processes are important to the development of consensus in science, there is a historical trend for scientists to become more concerned with complex multidisciplinary questions that require a deliberative approach. It therefore seems likely that deliberative consensus processes will become increasingly common across many areas of science.

Case Example 4.3: Human-Caused Global Warming

Scientific work on the effects of carbon dioxide in the atmosphere on the temperature of the Earth's surface goes back to the nineteenth century and continued to develop during the twentieth century (Rodhe et al., 1997). Growing concern about the role of human-caused emissions on global warming led the United Nations Environment Programme and the World Meteorological Organization to establish the Intergovernmental Panel on Climate Change (IPCC) in 1988. The IPCC uses a consensus process, with reports going through a series of steps, including governments and observer organizations nominating experts as potential authors, drafting of reports by the authors which are reviewed by a large number of experts, revision based on feedback and approval by all governments in the United Nations of the final documents (Intergovernmental Panel on Climate Change, 2023). The IPCC produced its first report in 1990, with subsequent reports in 1995, 2001, 2007, 2014 and 2023. Over successive reports, the consensus conclusions have become stronger. In 1990, the IPCC concluded that "The size of the warming over the last century is broadly consistent with the predictions of climate models but is also of the same magnitude as natural climate variability" (Houghton et al., 1990, p. xxix), whereas in 2023 it stated that "Human activities, principally through emissions of greenhouse gases, have unequivocally caused global warming, with global surface temperature reaching 1.1°C above 1850–1900 in 2011–2020" (Intergovernmental Panel on Climate Change, 2023, p. 42). The IPCC's consensus has been supported by statements from numerous national scientific organizations (Congressional Research Service, 2021). In addition, analysis of a random sample of climate-related scientific publications between 2012 and 2020 found greater than 99% consensus on human-caused global warming (Lynas et al., 2021), while surveys of scientists with relevant expertise showed 90–100% agreement (Cook et al., 2016). Nevertheless, there have been criticisms of the IPCC process. Curry and Webster (2013) have argued that the IPCC has downplayed the uncertainty of the data and that its consensus process is dominated by more confident scientists. There have also been dissenting groups, such as the Nongovernmental International Panel on Climate Change (Nongovernmental International Panel on Climate Change, 2023), which concluded that natural causes rather than human activity are the dominant cause of climate change, and the World Climate Declaration, which is a petition signed by a varied group of scientists who dispute a number of the IPCC's conclusions (Climate Intelligence, 2023). This area of science very clearly involves a deliberative consensus process. It involves a large body of evidence from several disciplines using a range of methodologies. The causality is complex and the quantification involves uncertainties, with the key issue being how big a contributor human activity is. Formal approaches to assessing consensus have played a key role.

4.3   Consensus and Dissensus

The case examples I have presented above all involve scientific questions where an eventual consensus was achieved. However, for many scientific questions (arguably for most), there is no current consensus. This is particularly true for emerging frontier areas where much scientific interest is focused. As Rescher (1993), one of the philosophical sceptics on scientific consensus, has noted:

> Throughout the areas near a research frontier there are always controversial issues that divide the community into conflicting and discordant schools of thought. And the rivalry between such schools is one of the main goads and incentives to the productive efforts of scientific researchers, each school being eager to validate its hunches and to vindicate its point of view. Dissensus is prevalent throughout science and provides one of the main stimuli to scientific progress. Far from science being a domain pervaded by consensus, there is, in fact, good reason to think that dissensus and controversy are the lifeblood of scientific work at and near the frontiers of research—though, to be sure, the rational and social dynamics of scientific opinion formation does generally make for an eventual uniformization of scientific opinion. Disagreement is in fact so pervasive and prominent a factor in science that it seems plausible to see dissensual debatability as a standard of value, and to regard a scientific question as important and interesting precisely in so far as there is room for disagreement about it. (p. 40)

Given the key role of dissensus at the frontiers of science, is there a role for consensus processes or are these a hindrance to progress? For scientific questions that involve complexity of evidence, I argue that there is value in assessing the degree of consensus because it can clarify areas of disagreement, examine characteristics of groups with differing views and be used to establish future research priorities. In effect, it can help better define the contested frontier and what needs to be done to advance in the area.

Case Example 4.4 illustrates this. It uses a deliberative consensus process to examine the factors responsible for the disappearance of Neanderthals. As well as clarifying areas of greater or lesser agreement, it explores whether the sociopolitical views of the scientists involved are influencing their beliefs.

Case Example 4.5 involves the lack of consensus around the biology of ageing and uses the results of an expert survey to suggest the sort of cross-disciplinary collaboration that is needed to progress the field.

Case Example 4.4: Factors Responsible for the Disappearance of Neanderthals

Neanderthals disappeared around 40,000 years ago, but the causes of this are highly contested. Numerous hypotheses have been put forward, which can be grouped into three categories. The first category relates the disappearance to the migration of modern humans into the territory occupied by Neanderthals and the resulting competition for limited resources. The second relates to the internal demographic dynamics of Neanderthal populations, such as whether they were too small in number to persist in the long run. The third category relates to environmental factors, such as climate changes or the introduction of pathogens by modern humans. To find out about whether there was an emerging consensus, Vaesen et al. (2021) conducted a survey of 216 palaeo-anthropologists, most of whom had published on Neanderthals. The experts were asked to rate 11 hypotheses as potential causal factors. The only one to be rated highly was "population size" (with a mean rating of 4.44 on a scale from 1 to 6). There was no consensus on competitive and environmental factors. The authors wondered whether experts' beliefs might be influenced by sociopolitical attitudes, such as endorsing anti-egalitarian values and the dominance of powerful groups, and the assigning of different moral worth to species other than modern humans (speciesism). They found no associations with these attitudes. Even though this is a topic where there is considerable dissensus, the consensus survey helped to progress the area by showing areas of greater or lesser consensus/dissensus and ruling out biases due to the scientists' social values.

Case Example 4.5: What Is Biological Ageing?

In 2019, an international symposium was held on the biology of ageing in Montreal, Canada. The symposium involved participants from a wide variety of disciplines with expertise related to ageing. The symposium included a debate on whether or not experts know what biological ageing is, which indicated a lack of consensus. Following the symposium, the organizers (Cohen et al., 2020) undertook a formal survey of 37 of the participants on the issues raised in the debate. They found a high level of agreement (86%) that ageing did not proceed uniformly across tissues and that ageing is heterogeneous and cannot be measured with a single metric. However, there was a lack of consensus on some of the most fundamental questions in the field, for example "whether we have a good understanding of the basic biological mechanisms of aging, whether it will soon be possible to reliably measure aging, whether there are many species that do not age appreciably, whether aging mechanisms are similar across species, whether aging is largely a cellular and molecular process, and whether aging is genetically programmed" (p. 6). The organizers concluded that this lack of consensus indicated the need to develop research links across a variety of subdisciplines involving mechanistic, evolutionary and demographic approaches to ageing.

These two case examples taken together also illustrate the possibility that there may never be a strong consensus on some topics. Research on the biology of ageing is virtually unlimited in its potential for further research and could one day conceivably lead to a consensus on basic mechanisms of ageing. By contrast, research on the demise of Neanderthals deals with past events where the possibilities of new evidence are limited and there may never be enough evidence to lead to any certainty.

The examples of deliberative consensus I have presented have been largely aimed at establishing what is true, but another major use of deliberative consensus is aimed at recommending what actions should be taken arising from the scientific evidence. This use of consensus is the focus of the next chapter.

References

Briant, J. (2005). Consensus can be wrong. *Institute of Public Affairs Review, 57*(Dec 2005), 35.

Catanzaro, M. (2024). Citation cartels help some mathematicians – And their universities – Climb the rankings. *Science, 383*(6682). https://doi.org/10.1126/science.zcl2s6d

Climate Intelligence. (2023). *World climate declaration: There is no climate emergency.* CLINTEL. Retrieved February 21, 2023, from https://clintel.org/world-climate-declaration/

Cohen, A. A., Kennedy, B. K., Anglas, U., Bronikowski, A. M., Deelen, J., Dufour, F., et al. (2020). Lack of consensus on an aging biology paradigm? A global survey reveals an agreement to disagree, and the need for an interdisciplinary framework. *Mechanisms of Ageing and Development, 191,* 111316. https://doi.org/10.1016/j.mad.2020.111316

Congressional Research Service. (2021). *Evolving assessments of human and natural contributions to climate change.* C. R. Service. https://www.everycrsreport.com/files/2021-08-11_R45086_3741fdd7111e5c017404ff7db8d8f5a90fcbe5a7.pdf

Cook, J., Oreskes, N., Doran, P. T., Anderegg, W. R. L., Verheggen, B., Maibach, E. W., et al. (2016). Consensus on consensus: A synthesis of consensus estimates on human-caused global warming. *Environmental Research Letters, 11*(4), 048002. https://doi.org/10.1088/1748-9326/11/4/048002

Curry, J. A., & Webster, P. J. (2013). Climate change: No consensus on consensus. *CAB Reviews, 8*(001). https://doi.org/10.1079/PAVSNNR20138001

Houghton, J. T., Jenkins, G. J., & Ephraums, J. J. (Eds.). (1990). *Climate change: The IPCC scientific assessment.* Cambridge University Press.

Intergovernmental Panel on Climate Change. (2023). *Climate change 2023: Synthesis report. Contribution of Working Groups I, II and III to the sixth assessment report of the Intergovernmental Panel on Climate Change.* IPCC.

Kanno, T., Iijima, K., Abe, Y., Koike, T., Shimada, N., Hoshi, T., et al. (2013). Peptic ulcers after the Great East Japan earthquake and tsunami: Possible existence of psychosocial stress ulcers in humans. *Journal of Gastroenterology, 48*(4), 483–490. https://doi.org/10.1007/s00535-012-0681-1

Kirshner, R. P. (2013). The accelerating universe: A Nobel surprise. *Proceedings of the American Philosophical Society, 157,* 438–456.

Lee, J., & O'Morain, C. (1997). Who should be treated for Helicobacter pylori infection? A review of consensus conferences and guidelines. *Gastroenterology, 113*(6 Suppl), S99–S106. https://doi.org/10.2307/4473279210.1016/S0016-5085(97)80021-2

Levenstein, S., Rosenstock, S., Jacobsen, R. K., & Jorgensen, T. (2015). Psychological stress increases risk for peptic ulcer, regardless of Helicobacter pylori infection or use of nonsteroidal anti-inflammatory drugs. *Clinical Gastroenterology and Hepatology,* *13*(3), 498–506 e491. https://doi. org/10.1016/j.cgh.2014.07.052

Lynas, M., Houlton, B. Z., & Perry, S. (2021). Greater than 99% consensus on human caused climate change in the peer-reviewed scientific literature. *Environmental Research Letters,* *16*(11), 114005. https://doi. org/10.2307/4473279210.1088/1748-9326/ac2966

Marshall, B. J., & Warren, J. R. (1984). Unidentified curved bacilli in the stomach of patients with gastritis and peptic ulceration. *Lancet, 1*(8390), 1311–1315. https://doi.org/10.2307/4473279210.1016/s0140-6736(84)91816-6

National Institutes of Health. (1994). NIH Consensus Conference. Helicobacter pylori in peptic ulcer disease. NIH consensus development panel on helicobacter pylori in peptic ulcer disease. *JAMA, 272*(1), 65–69.

Nongovernmental International Panel on Climate Change. (2023). *NIPCC Nongovernmental International Panel on Climate Change.* The Heartland Institute. Retrieved February 21, 2023, from http://climatechangereconsidered.org/

Perlmutter, S., Aldering, G., Goldhaber, G., Knop, R. A., Nugent, P., Castro, P. G., et al. (1999). Measurements of Ω and Λ from 42 high-redshift supernovae. *Astrophysical Journal, 517*(2), 565. https://doi.org/10.1086/307221

Pincock, S. (2005). Nobel Prize winners Robin Warren and Barry Marshall. *Lancet, 366*(9495), 1429. https://doi.org/10.1016/s0140-6736(05)67587-3

Rescher, N. (1993). *Pluralism: Against the demand for consensus.* Clarendon Press; Oxford University Press. Publisher description http://www.loc.gov/catdir/enhancements/fy0639/93018392-d.html

Riess, A. G., Filippenko, A. V., Challis, P., Cloccchiatti, A., Diercks, A., Garnavich, P. M., et al. (1998). Observational evidence from supernovae for an accelerating universe and a cosmological constant. *Astronomical Journal, 116*(3), 1009. https://doi.org/10.1086/300499

Rodhe, H., Charlson, R., & Crawford, E. (1997). Svante Arrhenius and the greenhouse effect. *Ambio, 26,* 2–5.

Vaesen, K., Dusseldorp, G. L., & Brandt, M. J. (2021). An emerging consensus in palaeoanthropology: Demography was the main factor responsible for the disappearance of Neanderthals. *Scientific Reports, 11*(1), 4925. https://doi. org/10.1038/s41598-021-84410-7

Vickers, P. (2023). *Identifying future-proof science.* Oxford University Press.

5

Expert Consensus to Guide Practice and Policy

The previous chapter showed how deliberative approaches to consensus are widely used for complex scientific questions where the evidence is extensive and also involves multiple disciplines and methodologies. In such cases, the individual scientist will generally not have the time or the technical expertise to critically read all the primary evidence. To come to a judgement on such questions, the consensus of a group of experts with diverse expertise is required.

This chapter looks at how deliberative consensus is also often used for making evidence-based recommendations to guide professional practice and public policy. For these decisions, the primary scientific evidence is again often complex and extensive, and it will be beyond the capabilities of practitioners and policymakers to critically process it all. Even to read systematic reviews of the evidence on a scientific question can be a challenge. In the area of medicine, for example, it has been estimated that in the year 2000 there were 1432 systematic reviews in the PubMed database, and that by 2019 this had increased to 29,073, with around 80 new systematic reviews published per day (Hoffmann et al., 2021). The historical trend indicates that this growing tide of research will become even more overwhelming in the future. As well as

© The Author(s) 2025
A. Jorm, *Expert Consensus in Science*, https://doi.org/10.1007/978-981-97-9222-1_5

this quantitative increase, there is also a trend for research studies and reviews to become more narrowly focused and specialized as interest in a topic grows, making it even harder for the practitioner or policymaker to make use of them.

However, deliberative consensus to guide practice and policy involves considerations additional to what the scientific evidence shows, in particular value judgements about various courses of action. The scientific evidence will not in itself indicate a clear course of action and must be considered in the light of various competing values which may differ from person to person or from interest group to interest group. These value judgements involve not only the scientists and other professionals with expertise in the area but also the values of individuals and groups who may be affected by any actions taken (a topic which is covered in more detail in Chap. 7). One of the reasons for the public controversy about issues like the role of human activity in climate change, the safety of GM food and the immunization of infants is that different values come into play. For example, even if it is agreed that the evidence shows a role for human activity in global warming, the actions that are taken as a result of this finding will depend on the weighing of various values. A person who is employed in the coal mining industry may place a greater weight on the risk to their employment and the economic decline of their community if mining is phased out than on the impact of sea level rise in Pacific-island countries or the loss of species due to global warming. Similarly, with immunization of infants, some parents may give greater weight to any possibility of long-term harm to their child than to potential benefits of preventing a rarely seen infectious disease.

The combining of scientific evidence with value judgements about different courses of action has been most fully developed in the area of medicine through the evidence-based medicine (EBM) movement and serves as an exemplar for how deliberative consensus could be used in other areas of professional practice and public policy.

5.1   Deliberative Consensus to Guide Medical Practice

Although the use of scientific evidence in medicine goes back centuries, the term *EBM* and the associated movement dates from the 1980s and 1990s. An early definition of the concept that is widely quoted comes from Sackett et al. (1996):

> Evidence based medicine is the conscientious, explicit, and judicious use of current best evidence in making decisions about the care of individual patients. By individual clinical expertise we mean the proficiency and judgment that individual clinicians acquire through clinical experience and clinical practice. Increased expertise is reflected in many ways, but especially in more effective and efficient diagnosis and in the more thoughtful identification and compassionate use of individual patients' predicaments, rights, and preferences in making clinical decisions about their care. (p. 71)

It is clear from this definition that clinical decisions are not directly determined by the evidence, but rather the evidence is taken into account by the clinician when making decisions about the care of a particular patient along with "predicaments, rights, and preferences".

Major activities of EBM are the carrying out of systematic reviews of the evidence on specific clinical questions and the development of clinical practice guidelines on how this evidence should be considered when making decisions about the care of patients.

When carrying out a systematic review, there is a thorough attempt to locate all the relevant evidence and to rate each study for its quality. A major vehicle for reviewing the evidence on health interventions is the Cochrane Collaboration (https://www.cochrane.org/), which is a UK-based organization with collaborating members from many countries who carry out systematic reviews according to a well-defined and rigorous methodology (Chandler & Hopewell, 2013). The Collaboration is named after Archie Cochrane, a Scottish medical practitioner who was an early advocate of testing medical interventions in randomized controlled trials. Another organization providing systematic reviews is JBI (Barker et al., 2023), which describes itself as "a global organisation promoting and supporting evidence-based decisions that improve health and health service delivery" (https://jbi.global/).

Early work on systematic reviews of health interventions often used a "hierarchy of evidence" to rate the quality of specific studies, with randomized controlled trial evidence at the top of the hierarchy. In some of these hierarchies, expert consensus is rated as a low level of evidence. This is the case with JDI's hierarchy of evidence (as described in Chap. 1), which is paradoxical because the methodological standards used by organizations like the Cochrane Collaboration and JBI are all developed by expert consensus. However, as described below, the methods for assessing the quality of evidence that are used in EBM have evolved over time to consider factors other than the research design.

The other major component of EBM is the production of clinical practice guidelines. The US Institute of Medicine (2011, p. 4) has defined these as "statements that include recommendations intended to optimize patient care that are informed by a systematic review of evidence and an assessment of the benefits and harms of alternative care options". The Institute states that for guidelines to be trustworthy, they should

- be based on a systematic review of the existing evidence;
- be developed by a knowledgeable, multidisciplinary panel of experts and representatives from key affected groups;
- consider important patient subgroups and patient preferences as appropriate;
- be based on an explicit and transparent process that minimizes distortions, biases and conflicts of interest;
- provide a clear explanation of the logical relationships between alternative care options and health outcomes, and provide ratings of both the quality of evidence and the strength of the recommendations; and
- be reconsidered and revised as appropriate when important new evidence warrants modifications of recommendations. (p. 5)

According to this definition, clinical practice guidelines involve a systematic review of the evidence, but go further to draw out the implications of the evidence for clinical practice taking into account the weighing of potential benefits and harms and the values of people with the health condition, namely "be developed by…representatives from key affected groups" and "consider… patient preferences".

Clinical practice guidelines are produced by a number of national organizations to a defined standard, and these standards are themselves consensus based. These organizations include the Institute for Quality and Efficiency in Health Care in Germany, National Institute for Health and Care Excellence (NICE) in the United Kingdom, the National Health and Medical Research Council in Australia and the Centers for Disease Control and Prevention in the United States. Clinical practice guidelines are also commonly produced by professional societies for their members. Case Example 5.1 describes the process used by NICE to develop guidelines.

One of the consequences of the use of hierarchies of evidence when producing clinical practice guidelines has been a distinction between "evidence-based guidelines" and "consensus-based guidelines". Guideline developers often refer to recommendations as "evidence-based" when there is moderate or high-quality evidence in a hierarchy, and "consensus-based" when there is low- or very low-quality evidence (Yao et al., 2021). However, this distinction is misleading, because all evidence requires interpretation by a group of experts for it to be translated into recommendations for

---

Case Example 5.1: How NICE Develops Clinical Practice Guidelines

NICE uses a seven-step process to develop its guidelines (NICE National Institute for Health and Care Excellence, 2023):

1. Topics for guidelines are referred to NICE by a number of organizations, including NHS England, Department of Health and Social Care and Department of Education.
2. A draft scope is written stating why the guideline is needed, what it will and will not cover and what it intends to achieve. The draft is provided to organizations with an interest in the topic for comment, and the scope is revised taking these into account.
3. The guideline is developed by systematically reviewing and summarizing the relevant evidence. The impact of the guideline on costs is also considered. The evidence is then considered by a committee consisting of practitioners and other professionals, and those who use services and their family members.
4. The draft guideline is sent to stakeholders for comment, and it is assessed for its impact on equality.
5. The comments are considered and the guideline is revised.
6. The guideline is signed off by a senior team at NICE and published.
7. The guidelines are updated regularly.

clinical practice. Even where randomized controlled trial evidence exists, it may be based on atypical patients who have only a single diagnosis rather than multi-morbidity, it may not reflect the operation of the health system of the country which the guidelines are designed for, and the trial participants may not be typical of the ethnic make-up of the population of interest. Experts have to make judgements about how applicable this evidence is to typical patients in the population of interest. Judgements also have to be made about how the costs and benefits of an intervention are weighed up. For example, should a very expensive new treatment which is marginally more effective than a cheap standard treatment be recommended? As Djulbegovic and Guyatt (2019) have argued:

> all clinical practice guideline recommendations, whether the available evidence is considered as being of high quality or very low quality, require both a judicious consideration of the relevant evidence and consensus from the panel regarding both the interpretation of the evidence and the tradeoff between the benefit vs the harm or burden of the recommended health intervention.
>
> As a result, making a distinction between evidence-based and consensus-based guidelines is both misguided and misleading because both require consensus. The crucial difference between evidence-based medicine and non–evidence-based medicine methods is that the former necessitates that judgments are consistent with underlying evidence, whereas the latter do not. (p. 726)

A more realistic hierarchy of evidence would then have "consensus based on randomized controlled trial evidence" at the top and "consensus based on clinical opinion" at the bottom.

In recent years, the rating of evidence has evolved away from hierarchies of evidence towards the more sophisticated Grading of Recommendations Assessment, Development and Evaluation (GRADE) system for assessing the strength of evidence and the strength of clinical practice recommendations (Guyatt et al., 2008). In GRADE, the evidence is rated from high quality (further research is very unlikely to change confidence in the estimate of effect) to very low quality (any estimate of effect is very uncertain). Evidence from randomized controlled trials is generally rated higher than evidence from observational studies, but the strength of evidence for any type of study can be rated up or down depending on factors such as study limitations, inconsistency of results, indirectness of evidence, imprecision of estimates of effect and reporting bias. The quality of evidence is then taken together with other factors to rate the strength of a recommendation for practice as "strong" or "weak" (sometimes termed "conditional"). The other factors involve matters of values, such as the balance between benefits and harms, patients' values and preferences, and whether the intervention represents a wise use of resources. Case Example 5.2 describes the use of GRADE to develop a clinical practice guideline on treatment of hypersomnolence.

Case Example 5.2: Development of a Clinical Practice Guideline on Treatment of Central Disorders of Hypersomnolence

The American Academy of Sleep Medicine has developed a range of clinical practice guidelines using the GRADE process, including one on the treatment of central disorders of hypersomnolence (severe daytime sleepiness) (Maski et al., 2021). To produce the guideline, the Academy commissioned a task force of clinicians with expertise on the topic. The experts had to declare all potential conflicts of interest. Those with high levels of conflict of interest were not allowed to be appointed, while those with lower levels had to recuse themselves from involvement in issues where that conflict was relevant. The task force carried out a systematic review of evidence on prescription medications and non-pharmacologic interventions. The clinical practice recommendations were then made according to the GRADE process. The task force determined the strength of each recommendation based on an overall assessment of the quality of evidence, balance of beneficial and harmful effects, patient values and preferences and resource use. A draft of the systematic review and the guideline were made available for public comment by Academy members and the general public, including patient advocacy groups. The task force considered all comments before producing a final version of the guideline. Recommendations were labelled as "strong", where almost all patients should receive the recommended action, or "conditional", where most patients should receive the course of action, but different choices may be appropriate to different patients depending on their values and preferences. Strong recommendations were worded as "We recommend that clinicians....", while conditional recommendations used the words "We suggest that clinicians...". The guideline qualifies its recommendations with the following advice: "The ultimate judgment regarding any specific treatment must be made by the treating clinician and the patient, taking into consideration the individual circumstances of the patient, available treatment options, and resources". It is planned that the guideline will be updated as further research becomes available.

Implementing the GRADE system involves some quite complex judgements for the guideline development expert panel. To assist with these judgements, the developers of GRADE (the GRADE Working Group) have gone on to produce an "Evidence to Decision" framework which aims to make decision-making more systematic and transparent (Alonso-Coello et al., 2016). The use of this framework requires the guideline development panel to choose a values perspective, which may be individual patient, population or health system, as the values that apply will vary according to the perspective taken. For example, if the guideline involves clinical recommendations from an individual patient perspective, the panel needs to consider the following:

- The priority of the problem. Is the problem a priority?
- Benefits and harms. How substantial are the desirable and undesirable anticipated effects?
- Certainty of the evidence. What is the overall certainty of the evidence of effects?
- Outcome importance. Is there important uncertainty or variability in how much people value the outcome?
- Balance. Does the balance of desirable or undesirable effects favour the intervention compared to an alternative?
- Resource use. Is the intervention cost-effective for the individual?
- Equity. What is the impact of the intervention on health equity?
- Acceptability. Is the intervention acceptable to patients, families and healthcare providers?
- Feasibility. Is the intervention feasible for patients, families and healthcare providers?

If the perspective adopted was from the population and health system, rather than the individual patient, then the question about resource use would consider cost-effectiveness for the society as a whole, and acceptability and feasibility would be considered for key stakeholders in the society.

The evolution of guideline development from a relatively simple "Levels of Evidence" approach to a more sophisticated "Evidence to

Decision" framework involves much greater complexity in decision-making, but the GRADE Working Group (Alonso-Coello et al., 2016) argues that it provides

> an approach to structured reflection that can help those making recommendations or decisions to be more systematic and explicit about the judgments that they make, the evidence used to inform each of those judgments, additional considerations, and the basis for their recommendations or decisions. (p. 9)

The EBM movement has had many critics along the way, but has adapted its approaches to deal with some of the early criticisms, such as taking more explicit account of the importance of values. However, there have been persisting criticisms about the role of industry, particularly the pharmaceutical industry, as a funder of much of the evidence from randomized controlled trials. Jureidini and McHenry (2022), for example, have charged that the financial interests of industry have trumped the common good, with the funders of trials able to suppress negative trial results, failing to report adverse events and withholding raw data from independent analysis. Another critic, a Scottish general practitioner, has pulled no punches in a viewpoint article entitled "Evidence based medicine is broken", which was published in the prestigious medical journal *BMJ* (Spence, 2014):

> You see, without so called "evidence" there is no seat at the guideline table. This is the fundamental "commissioning bias," the elephant in the room, because the drug industry controls and funds most research. So the drug industry and EBM have set about legitimising illegitimate diagnoses and then widening drug indications, and now doctors can prescribe a pill for every ill... How many people care that the research pond is polluted, with, fraud, sham diagnosis, short term data, poor regulation, surrogate ends, questionnaires that can't be validated, and statistically significant but clinically irrelevant outcomes? Medical experts who should be providing oversight are on the take. Even the National Institute for Health and Care Excellence and the Cochrane Collaboration do not exclude authors with

conflicts of interest, who therefore have predetermined agendas. The current incarnation of EBM is corrupted, let down by academics and regulators alike. (p. 1)

There is data indicating that conflicts of interest can indeed distort expert panel judgements. Nejstgaard et al. (2020) examined a large number of clinical guidelines, advisory committee reports, opinion pieces and review articles to see if there was any association between conflicts of interest and favourable recommendations. They did find an association between financial conflicts of interest (industry funding of the documents and authors' company ties) and favourable recommendations on drugs and devices in these documents.

One approach is to exclude experts with potential conflicts of interest. However, excluding such experts can itself be problematic. People with the highest level of expertise might well have industry collaborations. Even where an expert has no industry links, they may have carried out considerable research on a particular treatment, believe strongly in that treatment and want to see if promoted.

Another solution is to engage a broader range of stakeholders in prioritizing what research is carried out. One initiative of this kind is the James Lind Alliance which is a UK organization that facilitates the identification of clinical research priorities shared by clinicians, patients and carers (Chalmers et al., 2013). Work carried out by the Alliance has shown that there is often a mismatch between what research is being done and what patients and clinicians want to see carried out. Existing commercial trials were found to largely evaluate drugs, vaccines and biologicals, whereas patients and clinicians wanted to see a greater emphasis on education and training, service delivery, psychological therapy, physical therapies, complementary therapies, social care and diet. To overcome this gap, the Alliance has set up clinician–patient priority setting partnerships for a wide range of diseases. The resulting highest priorities are provided to research funders.

The take-away message here is that deliberative consensus decisions are limited by the availability and quality of research data that underpins experts' judgements and by biases in evaluating the evidence due to conflicts of interest that arise through association with industry or by strong commitments arising from the expert's own research findings. However, engaging a broader range of stakeholders with diverse values in decisions about what research is done can help to reduce these limitations.

Despite the criticisms, the EBM movement has had considerable influence on healthcare. In 2007, the medical journal *BMJ* asked readers to vote on the importance of 15 short-listed medical milestones since 1840. The number 1 milestone was sanitation, with EBM coming in number 8, ahead of advances such as medical imaging, computers, immunology and the risks of smoking (Ferriman, 2007).

5.2 Deliberative Consensus in the Development of Health Policies

The EBM movement has been largely concerned with guiding the decisions that health practitioners make about the care of individual patients. However, deliberative consensus approaches are also used to guide health policy, where the target is a whole population rather than an individual patient. Examples include addition of fluoride to drinking water, requirements for infant immunization, dietary recommendations on consumption of fruit and vegetables, mandatory supplementation of food with micronutrients, recommendations on avoidance of sun exposure and advice on reducing health risks from drinking alcohol. In these cases, the values considerations are different in that they concern what is important to the population as a whole, which may sometimes be at odds with what is important to individuals or sub-groups within that population. Decisions also become binary—either the policy is implemented or not—rather than on a continuum of strength of recommendation from weak to strong. Case Example 5.3 illustrates the considerations involved in one such universal health policy.

Case Example 5.3: Preventing Birth Defects by Universal Mandatory Food Fortification with Folic Acid

Maternal intake of folic acid starting before pregnancy is known to prevent most cases of infant spina bifida and anencephaly. Prescribing folic acid supplements to pregnant women is only a partial solution, as this can only be done once a pregnancy is discovered, which may be too late to prevent these defects. It has therefore been proposed that there be large-scale fortification of staple foods (e.g. wheat flour, maize flour and rice). The World Health Organization and the Food and Agriculture Organization of the United Nations (Allen et al., 2006) have released guidelines on food fortification with micronutrients, including folic acid. The guidelines were produced by a multidisciplinary panel of experts, covering expertise in public health, nutritional sciences and food technology, from both the public and private sectors. A draft of the guidelines was circulated to field nutritionists and public health practitioners and also tested in a number of countries. Comments received were taken into account in producing the final guidelines. While a number of countries have implemented mandatory food fortification with folic acid, most have not, leading to a recent urgent call for a World Health Assembly resolution on mandatory fortification, which it is hoped would encourage other governments to take action (Kancherla et al., 2022). Mandatory fortification has the advantage of not requiring active behavioural change (such as taking supplements). However, there is an ethical dilemma here in weighing up the benefits and risks in that everyone is exposed to any risks, but only a sub-group of the population benefits. As has been pointed out by Harvey and Diug (2018): "The aim of folic acid fortification…is to compensate a presumed genetic defect in individuals who are at risk but cannot be individually identified…Folic acid fortification thus raises ethical questions about exposing the many for the benefit of the few; concerns have been expressed that exposing children to high levels of folic acid over their lifetime may increase their risk of adverse effects" (p. 111).

5.3   Guidelines and Position Statements on Practice and Policy in Other Areas

The use of deliberative consensus processes to guide science-based practice and policy has been most prominent in the area of medicine, but is also seen in a variety of other areas. Paralleling the work of the Cochrane Collaboration in carrying out systematic reviews of evidence in medicine, there is the Campbell Collaboration (https://www.campbellcollaboration. org/), which carries out systematic reviews in social policy areas, including ageing, business and management, children's and young persons' well-being, climate solutions, crime and justice, disability, education, international development and social welfare. The Collaboration also develops methods for systematic reviewing and for knowledge translation and implementation (Chandler et al., 2017).

As in medicine, systematic reviews provide only part of the input for practice and policy, and need to be combined with a consensus on values. Arguably, the most visible example of the use of consensus to develop policy outside of medicine is by the IPCC in relation to the role of human activity in global warming. The reports produced by the IPCC both synthesize the scientific evidence and make recommendations for policymakers. These two functions of the reports are intertwined, but it is often not clear when a statement of scientific fact is being made and when it is a matter of value judgement. Furthermore, where value judgements are made, it is not clear how these were arrived at. Take, for example, the following two conclusions from the sixth IPCC report, both of which are made with "high confidence" (Core Writing Team et al., 2023):

A.1.1 Global surface temperature was 1.09 [0.95 to 1.20]°C higher in 2011–2020 than 1850–1900, with larger increases over land (1.59 [1.34 to 1.83]°C) than over the ocean (0.88 [0.68 to 1.01]°C). Global surface temperature in the first two decades of the 21st century (2001–2020) was 0.99 [0.84 to 1.10]°C higher than 1850–1900. Global surface temperature has increased faster since 1970 than in any other 50-year period over at least the last 2000 years (high confidence). (p. 4)

C.5 Prioritising equity, climate justice, social justice, inclusion and just transition processes can enable adaptation and ambitious mitigation actions and climate resilient development. Adaptation outcomes are enhanced by increased support to regions and people with the highest vulnerability to climatic hazards. Integrating climate adaptation into social protection programs improves resilience. Many options are available for reducing emission-intensive consumption, including through behavioural and lifestyle changes, with co-benefits for societal well-being. (high confidence). (p. 31)

The former is a statement about the best estimate that can be made on surface temperature change and does not call for any specific action to be taken. By contrast, the latter is a statement of the values that the authors place on "equity, climate justice, social justice, inclusion and just transition processes" and calls for various policy changes, such as "increased support to regions and people with the highest vulnerability", "social protection programs" and "behavioural and lifestyle changes". The fact that both statements are rated "high confidence" further blurs the distinction between claims about scientific facts and calls for action based on applying values to these facts.

Two additional examples help to illustrate the complex interplay between scientific knowledge and values in deliberative consensus on professional practice and policy. The first involves the development of professional practice standards in forensics by the American Academy of Forensic Sciences (2024) (see Case Example 5.4). In these standards, procedures are specified for collecting and examining evidence, but the statement of underlying values is rather general. Standards documents are prefaced by a statement that the vision of the Standards Board is to safeguard "Justice, Integrity and Fairness" and that the Board values "integrity, scientific rigour, openness, due process, collaboration, excellence, diversity and inclusion". However, the more specific values behind standards are implicit. Presumably, the more specific aim of forensic standards is to help judges and juries to more accurately determine guilt or innocence. Such judgements mean that the forensic method has to reduce both false positives (people judged as guilty who are really innocent) and false negatives (people judged as innocent who are really guilty). One common dictum of law in English-speaking jurisdictions is "Blackstone's ratio", according to which "It is better that ten guilty persons escape than that one innocent suffer", which implies that false positives are more serious errors than false negatives. However, in the case of the AAFS standards, such underpinning values are not directly addressed.

Case Example 5.4: American Academy of Forensic Sciences (AAFS) Standards

The AAFS has a Standards Board which provides consensus-based forensic standards within an accredited framework from the American National Standards Institute (American Academy of Forensic Sciences, 2024). Existing standards cover a wide variety of areas of practice, such as examination and documentation of footwear and tire impression evidence, collection of known DNA samples from domestic animals, examination of handwritten items, preservation and examination of charred documents, and resolving comingled remains in forensic anthropology. The Standards Board selects and removes members of Consensus Bodies which produce specific standards. A Consensus Body has 7–25 members who are all volunteers. In appointing people to a Consensus Body, the Board ensures that no single interest category dominates. Interest categories include academics and researchers, jurisprudence and criminal justice, producers of forensic products, and government and non-government entities involved in forensic work. The Consensus Bodies produce draft standards that are put out for public comment and revised in response to the comments received. Members of Consensus Bodies vote on the final version of a standard, with a two-thirds majority required for approval. Standards produced by a Consensus Body are then voted on by the Standards Board, with majority approval required. Standards are regularly revised.

The second example involves the development of the Sustainable Development Goals by the United Nations Development Programme (see Case Example 5.5). What is different about this process is that it largely involved coming to a consensus on the Goals, which are a matter of values. Science-based knowledge came in subsequently as a means to fulfill Goals, again using a consensus approach. Taken together, these two examples show how either a consensus on scientific knowledge or a consensus on values can take the dominant position.

Given that both scientific knowledge and values are involved in deliberative consensus on professional practice and policy, the question arises as to who provides the expertise on values. Scientists and science-informed professionals provide expertise on the knowledge, but they are not necessarily the most appropriate people to inform the values. I turn to this topic in Chap. 7. However, before doing so, I consider another area of science in which consensus plays a role, which is determining what are acceptable research methods.

Case Example 5.5: The United Nations Development Programme's Sustainable Development Goals

In 2012, at the United Nations Conference on Sustainable Development, there was agreement from member states to develop a set of Sustainable Development Goals (United Nations Department of Economic and Social Affairs, 2024). A 30-member Open Working Group was given the task of developing these Goals. The Open Working Group was charged with adopting its recommendations by consensus, reflecting different options if necessary. The Group could draw on the support of a technical support team and expert panels. However, the exact procedure for arriving at the Goals is not clear. Based on the work of the Open Working Group, in 2015, the United Nations committed to 17 interlinked Sustainable Development Goals. These are: no poverty; zero hunger; good health and well-being; quality education; gender equality; clean water and sanitation; affordable and clean energy; decent work and economic growth; industry, innovation and infrastructure; reduced inequalities; sustainable cities and communities; responsible consumption and production; climate action; life below water; life on land; peace, justice and strong institutions; and partnerships for the goals. The goals have 169 associated targets and 232 indicators.

Because science has a potentially important role in the implementation of the Goals, in 2016 the United Nations appointed a 15-member Independent Group of Scientists representing a variety of scientific disciplines and institutions to report on how science can contribute to sustainable development. The Group is mandated to report every four years, with its first report published in 2019 (Independent Group of Scientists appointed by the secretary-general, 2019). One of the proposals in the report was for the scale-up of what it called "sustainability science", which was defined as an "academic field of studies that sheds light on complex, often contentious and value-laden, nature-society interactions, while generating usable scientific knowledge for sustainable development" (p. 120). As an example of sustainability science, the report cited an example of the phasing out of coal in Europe: "There was found to be less resistance in the coal-mining regions where scientists, policymakers, and coal miners had come together to jointly identify alternatives for regional development and individual livelihoods" (p. 120).

References

Allen, L., de Benoist, B., Dary, O., & Hurrell, R. (Eds.). (2006). *Guidelines on food fortification with micronutrients*. World Health Organization and Food and Agriculture Organization of the United Nations. https://www.who.int/publications/i/item/9241594012

Alonso-Coello, P., Oxman, A. D., Moberg, J., Brignardello-Petersen, R., Akl, E. A., Davoli, M., et al. (2016). GRADE Evidence to Decision (EtD) frameworks: A systematic and transparent approach to making well informed healthcare choices. 2: Clinical practice guidelines. *BMJ, 353*, i2089. https://doi.org/10.1136/bmj.i2089

American Academy of Forensic Sciences. (2024). *Academy Standards Board*. Retrieved August 16, 2024, from https://www.aafs.org/academy-standards-board/about-asb

Barker, T. H., Stone, J. C., Sears, K., Klugar, M., Leonardi-Bee, J., Tufanaru, C., et al. (2023). Revising the JBI quantitative critical appraisal tools to improve their applicability: An overview of methods and the development process. *JBI Evidence Synthesis, 21*(3), 478–493. https://doi.org/10.11124/JBIES-22-00125

Chalmers, I., Atkinson, P., Fenton, M., Firkins, L., Crowe, S., & Cowan, K. (2013). Tackling treatment uncertainties together: The evolution of the James Lind Initiative, 2003-2013. *Journal of the Royal Society of Medicine, 106*(12), 482–491. https://doi.org/10.1177/0141076813493063

Chandler, J., Churchill, R., Higgins, J. P. T., Lasserson, T., & Tovey, D. (2017). *Methodological expectations of Campbell Collaboration intervention reviews (MECCIR): Reporting standards*. https://campbellcollaboration.org/meccir.html

Chandler, J., & Hopewell, S. (2013). Cochrane methods – Twenty years experience in developing systematic review methods. *Systematic Reviews, 2*, 76. https://doi.org/10.1186/2046-4053-2-76

Core Writing Team, Lee, H., & Romero, J. (Eds.). (2023). *IPCC, 2023: Climate change 2023: Synthesis report*. IPCC. https://doi.org/10.59327/IPCC/AR6-9789291691647

Djulbegovic, B., & Guyatt, G. (2019). Evidence vs consensus in clinical practice guidelines. *JAMA, 322*(8), 725–726. https://doi.org/10.1001/jama.2019.9751

Ferriman, A. (2007). BMJ readers choose the "sanitary revolution" as greatest medical advance since 1840. *BMJ, 334*, 111. https://doi.org/10.1136/bmj.39097.611806.DB

Guyatt, G. H., Oxman, A. D., Vist, G. E., Kunz, R., Falck-Ytter, Y., Alonso-Coello, P., et al. (2008). GRADE: An emerging consensus on rating quality of evidence and strength of recommendations. *BMJ, 336*(7650), 924–926. https://doi.org/10.1136/bmj.39489.470347.AD

Harvey, K. J., & Diug, B. O. (2018). The value of food fortification as a public health intervention. *Medical Journal of Australia, 208*(3), 111–112. https://doi.org/10.5694/mja17.01095

Hoffmann, F., Allers, K., Rombey, T., Helbach, J., Hoffmann, A., Mathes, T., & Pieper, D. (2021). Nearly 80 systematic reviews were published each day: Observational study on trends in epidemiology and reporting over the years 2000–2019. *Journal of Clinical Epidemiology, 138*, 1–11. https://doi.org/10.1016/j.jclinepi.2021.05.022

Independent Group of Scientists appointed by the Secretary-General. (2019). *Global sustainable development report 2019: The future is now – Science for achieving sustainable development.* United Nations.

Institute of Medicine (U.S.). Committee on Standards for Developing Trustworthy Clinical Practice Guidelines, & Graham, R. (2011). *Clinical practice guidelines we can trust.* National Academies Press.

Jureidini, J., & McHenry, L. B. (2022). The illusion of evidence based medicine. *BMJ, 376*, o702. https://doi.org/10.1136/bmj.o702

Kancherla, V., Botto, L. D., Rowe, L. A., Shlobin, N. A., Caceres, A., Arynchyna-Smith, A., et al. (2022). Preventing birth defects, saving lives, and promoting health equity: An urgent call to action for universal mandatory food fortification with folic acid. *Lancet Global Health, 10*(7), e1053–e1057. https://doi.org/10.1016/S2214-109X(22)00213-3

Maski, K., Trotti, L. M., Kotagal, S., Robert Auger, R., Rowley, J. A., Hashmi, S. D., & Watson, N. F. (2021). Treatment of central disorders of hypersomnolence: An American Academy of Sleep Medicine clinical practice guideline. *Journal of Clinical Sleep Medicine, 17*(9), 1881–1893. https://doi.org/10.5664/jcsm.9328

Nejstgaard, C. H., Bero, L., Hrobjartsson, A., Jorgensen, A. W., Jorgensen, K. J., Le, M., & Lundh, A. (2020). Association between conflicts of interest and favourable recommendations in clinical guidelines, advisory committee reports, opinion pieces, and narrative reviews: Systematic review. *BMJ, 371*, m4234. https://doi.org/10.1136/bmj.m4234

NICE National Institute for Health and Care Excellence. (2023). *How we develop NICE guidelines.* https://www.nice.org.uk/about/what-we-do/our-programmes/nice-guidance/nice-guidelines/how-we-develop-nice-guidelines

Sackett, D. L., Rosenberg, W. M., Gray, J. A., Haynes, R. B., & Richardson, W. S. (1996). Evidence based medicine: What it is and what it isn't. *BMJ, 312*(7023), 71–72. https://doi.org/10.1136/bmj.312.7023.71

Spence, D. (2014). Evidence based medicine is broken. *BMJ, 348*, g22. https://doi.org/10.1136/bmj.g22

United Nations Department of Economic and Social Affairs. (2024). *Sustainable development: The 17 goals.* Retrieved August 16, 2024, from https://sdgs.un.org/goals

Yao, L., Ahmed, M. M., Guyatt, G. H., Yan, P., Hui, X., Wang, Q., et al. (2021). Discordant and inappropriate discordant recommendations in consensus and evidence based guidelines: Empirical analysis. *BMJ, 375*, e066045. https://doi.org/10.1136/bmj-2021-066045

6

Expert Consensus on Research Methods

Previous chapters have discussed how spontaneous and deliberative consensus processes operate in establishing scientific truths (Chaps. 3 and 4), and how deliberative approaches have also been applied to develop consensus on professional practice and policy (Chap. 5). The present chapter examines a third area where consensus processes are integral—agreement among scientists about what research methods in their field are sound.

The term "method" can be used in different ways. To a philosopher of science, it might cover the "scientific method" in a broad sense as an empirical method for acquiring knowledge that is common across scientific disciplines. However, in this chapter, the term "research methods" is used in a more specific sense to cover methods for collecting, analysing, reporting, understanding and interpreting data in research studies.

When scientists carry out a research project, they want other scientists in their field to accept their findings. Acceptance of their findings is indicated by positive peer review of papers submitted for publication, and by subsequent positive citations of their work and an absence of negative citations. Acceptance of findings is more likely to occur if the scientists have carried out their research using methods that are widely accepted by their peers as best practice.

© The Author(s) 2025
A. Jorm, *Expert Consensus in Science*, https://doi.org/10.1007/978-981-97-9222-1_6

6.1   Spontaneous and Deliberative Processes with Research Methods

How does a scientist know that a particular research method will be accepted by peers? As for establishing scientific truths, consensus about research methods may arise spontaneously in the marketplace for sound methodologies or through a planned deliberative process. Table 6.1 summarizes the difference between these two processes when used for developing a consensus on research methods. Examples of these two processes are given later in this chapter.

Many methodological innovations are accepted (and sometimes rejected) by a spontaneous consensus. These are typically in areas where there is no existing method (e.g. a new laboratory or statistical method) or the existing methods have limitations and can be improved, so the level of innovation in the method is high. Such innovations are often proposed by a small team of scientists and made available for use by other scientists through a publication, patent, piece of equipment or software. These then enter a marketplace for valid methodologies and compete for adoption by other scientists in the field. A method can flourish and be

Table 6.1 Contrast between spontaneous and deliberative processes for developing consensus on research methods

Characteristic	Spontaneous consensus process	Deliberative consensus process
How consensus builds	Spontaneous adoption by peers, indicated by positive citations	A formal consensus process used to develop method
Nature of the methods gap to be filled	There is no existing method, or existing methods can be improved	Methods exist, but these need standardization or infrastructure for dissemination
Level of innovation in method	High	Lower
Complexity of providing and implementing the method	Often done by a small team	Requires coordination of efforts across a larger number of scientists; may require creating a new organization

widely adopted by peers, or it can die through neglect or peer criticism. For a method to flourish, it needs to be accepted as sound by experts in the field, fill a gap where no method previously existed or provide a superior alternative to existing methods. A scientist acknowledges that they have adopted a method by a positive citation of the source of that method. Positive citations are an indirect indicator of acceptance that a methodology is regarded as valid by researchers in the area. Spontaneously accepted methods often have very high counts of positive citations.

Other methodological innovations achieve consensus through a deliberative process. This more commonly occurs where there are existing methods, but these are not standardized, making comparison across research studies difficult. For example, researchers may use different measures or report their findings in different ways. The level of innovation in these methods is generally lower than with methods adopted through spontaneous consensus. These methods are complex to provide to other researchers or to implement, requiring coordination of efforts across a large group of scientists. In some cases, the organizational infrastructure to implement these methods is lacking and has to be developed. To deal with this complexity and the need to coordinate efforts across many scientists, formal consensus methods are used to develop agreement on the details of the method.

While Table 6.1 presents spontaneous and deliberative processes as separate for purposes of exposition, they are really ends of a continuum, as is the case for establishing scientific truths. For some research methods, there will be elements of both processes in establishing a consensus on their soundness.

6.2   Case Examples of Spontaneous Consensus

Three diverse case examples of spontaneous adoption of research methods are presented below. The first example (Case Example 6.1) concerns the use of statistical significance testing and, in particular, the widespread adoption of the convention that $P < .05$ is the defining threshold for significance.

Case Example 6.1 The *P* < .05 Level of Statistical Significance

Statistical significance tests are used widely in many areas of science. These tests give the probability that an effect would occur if the null hypothesis was true. A number of statisticians contributed to the development of statistical significance testing, but the statistician who made it popular with scientists was Ronald Fisher (1925) through his book *Statistical Methods for Research Workers*. Fisher wrote that the aim of this book was "to put into the hands of research workers, and especially biologists, the means of applying statistical tests accurately to numerical data" (p. 16). He suggested the *P* < .05 level for defining an effect as statistically significant: "The value for which P=.05, or 1 in 20, is 1.96 or nearly 2; it is convenient to take this point as a limit in judging whether a deviation is to be considered significant or not. Deviations exceeding twice the standard deviation are thus formally regarded as significant" (p. 47). What popularized Fisher's proposal for *P* < .05 is that his book provided user-friendly tables based on this value, at a time when researchers did not have computers to do the calculations required for exact *P*-values. Subsequently, Fisher and Yates (1938) published *Statistical Tables for Biological, Agricultural and Medical Research*, which further supported the ease of using the *P* < .05 decision rule. Since Fisher's time, there has been frequent criticism of statistical significance testing and the use of *P* < .05, and a number of alternative approaches have been proposed (Kennedy-Shaffer, 2019). These have included such approaches as lowering the significance level to *P* < .005, carrying out corrections for multiple statistical testing, using confidence intervals instead of *P*-values, putting more emphasis on effect sizes, and using Bayesian statistical methods. However, none of these alternatives have managed to displace statistical significance testing, and the *P* < .05 decision rule is still widely used. In fact, some analyses of the literature have shown that the use of statistical significance testing has increased in recent decades in both biomedical and general science journals (Chavalarias et al., 2016; Cristea & Ioannidis, 2018).

This example illustrates many characteristics typical of spontaneous consensus. There was no previous simple rule for making decisions on the significance of statistical data. Although there was prior work on statistical significance testing, Fisher consolidated this and made it accessible to the average scientist. He was able to do this with his own resources through publication of his book and statistical tables. His work was rapidly adopted and continues to be widely used.

The second example (Case Example 6.2) concerns an atlas of the rat brain, which was developed by two neuroscientists and became a standard reference work for researchers using the rat as an experimental model.

Case Example 6.2 Atlas of the Rat Brain

Rats are often used as a model for studying human diseases, including for diseases of the brain like Alzheimer's and Parkinson's. In order to study the rat brain, researchers need an accurate way to identify specific brain structures that may be affected by disease or injury. To facilitate this, George Paxinos and Charles Watson published their book *The Rat Brain in Stereotaxic Coordinates* in 1982, which was the first accurate stereotaxic (three-dimensional) atlas (Paxinos & Watson, 1982). The atlas was based on thin sections of a single rat's brain, which were shown with detailed photographs and drawings. It allowed researchers to use reference skull landmarks and make a small hole in the skull to locate a brain region where they could place injections or electrodes. The atlas was subsequently updated and improved and went through seven editions. According to Google Scholar, the atlas had been cited over 90,000 times by 2023. Figure 6.1 shows the number of citations per year following publication. There was a rapid and continuing rise over three decades, with some decline in recent years which may reflect the availability of alternative atlases.

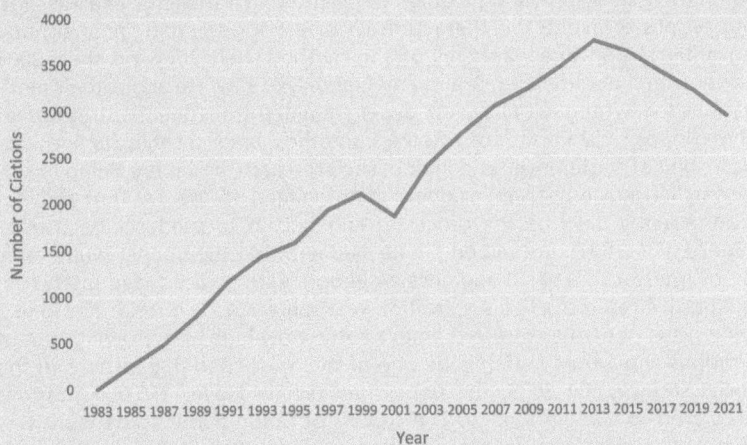

Fig. 6.1   Citations per year in Google Scholar to *The Rat Brain in Stereotaxic Coordinates*

This example is also typical of spontaneous consensus. Paxinos and Watson provided a superior method of locating areas of the rat's brain than was previously available. The two of them were able to do the required anatomical work themselves and publish their atlas in multiple editions. The atlas has been widely adopted and cited by neuroscientists who recognized its advantages over available alternatives.

The third example (Case Example 6.3) comes from the social sciences. It involves a key publication that defined a research method—thematic analysis—that was often used but poorly defined, and provided a usable "how to" guide for researchers wishing to use it.

Case Example 6.3 Thematic Analysis in Qualitative Research

Qualitative research methods are often used in the social sciences to analyse verbal data from interviews and focus groups. There are a number of different approaches to analysing such data, with thematic analysis being one of the most common. A seminal paper on thematic analysis was published by Virginia Braun and Victoria Clarke (2006). They defined thematic analysis as "a method for identifying, analysing and reporting patterns (themes) within data" (p. 79). They noted that while thematic analysis is widely used, "there is no clear agreement about what thematic analysis is and how you go about doing it" (p. 79). Braun and Clarke's paper attempted to fill this gap by providing a guide to performing a thematic analysis, noting pitfalls to avoid, the characteristics of a good thematic analysis, and advantages and disadvantages of the method. Reflecting on their 2006 paper over a decade later, Braun and Clarke (2019) wrote about their motivations for writing it: "Our 2006 paper stemmed from dual frustrations: at the 'sloppy mishmash'...of theories, method and techniques we saw described at conferences and in published research; and there being lots of research (from ourselves included) that claimed to 'do TA', but did not transparently describe the processes engaged in to produce the themes reported" (p. 591). Braun and Clarke assumed that their paper would only be of interest to a small audience of people with a specialized interest in qualitative research. This assumption was spectacularly wrong. As Fig. 6.2 shows, the paper has received an extraordinarily high and growing rate of citations and would certainly be one of the most cited publications in the social sciences. By 2023, the paper had received over 170,000 citations according to Google Scholar. The success of Braun and Clarke's work was due to the clear and practical guidance they provided about a research method that had previously been poorly defined and developed.

(continued)

Case Example 6.3 (continued)

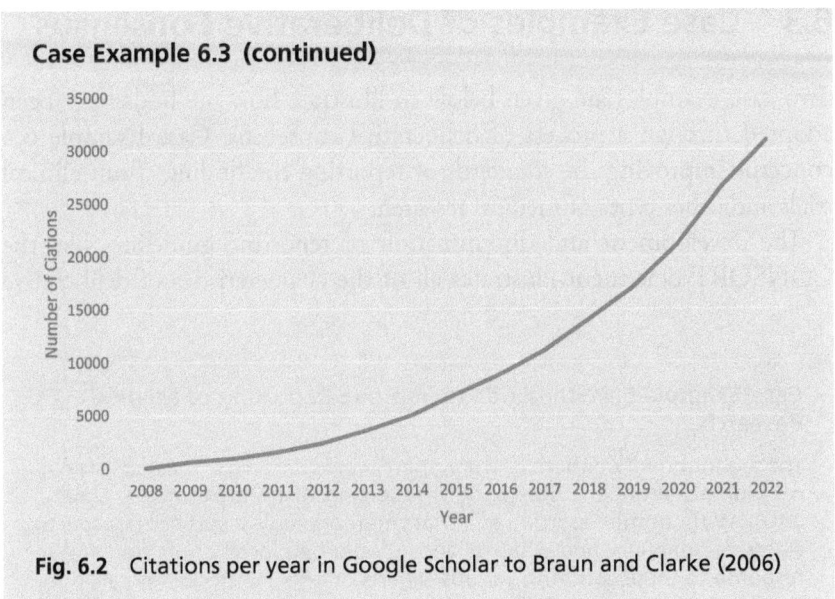

Fig. 6.2 Citations per year in Google Scholar to Braun and Clarke (2006)

Again, this example shows the characteristics of spontaneous consensus as summarized in Table 6.1. In their article on thematic analysis, Braun and Clarke provided a clearer description and guide to implementing this type of analysis than previous authors. The two of them were able to write and publish an article with the resources available to them, and it was spontaneously adopted by their peers for analysing qualitative data.

There are many other examples of spontaneous adoption of research methods that could be mentioned. These include Hill's (1965) criteria for using epidemiological evidence to infer that a risk factor is a cause of disease, Baron and Kenny's (1986) development of mediation analysis to infer that a variable provides a mechanism linking an independent variable to a dependent variable, Libby's (1946) development of radiocarbon dating of organic materials, and Mullis et al.'s (1986) discovery of how the polymerase chain reaction could be used to amplify DNA. All of these filled a clear gap with a novel approach, one scientist or a small team was able to develop and promulgate the method, and it was widely adopted and cited by researchers in the field.

6.3   Case Examples of Deliberative Consensus

Two case examples are given below to illustrate how methods have been adopted through a process of deliberative consensus. Case Example 6.4 concerns improving the standards of reporting the findings from clinical trials and other types of medical research.

The development and dissemination of reporting guidelines like the CONSORT Statement illustrates all of the characteristics of deliberative

---

Case Example 6.4   Standards to Improve Reporting of Medical Research

Throughout the twentieth century, there were concerns expressed in medical journals about the adequacy of reporting of trials (Altman & Simera, 2016). Without full reporting of a trial's methodology, it was not possible to assess its adequacy and what reliance should be placed on its findings. In response to these concerns, various experts on trial methodology called for the development of guidelines on best practice in reporting. An important milestone in improving reporting was the publication of the Consolidated Standards of Reporting Trials (CONSORT) Statement in 1996, which gave a checklist and flow diagram for reporting the results of randomized controlled trials. There is some evidence that the CONSORT Statement did improve the quality of reporting of trials, although reporting remained sub-optimal (Turner et al., 2012). The CONSORT recommendations have been updated a number of times since 1996, most recently in 2022. A multistage expert consensus process was used to develop the 2022 statement (Butcher et al., 2022). Firstly, potential outcome reporting items were generated by consultation with experts and a review of existing guidance on reporting trial outcomes. Next, there was a three-round international Delphi voting process involving 124 panelists from 22 countries. Finally, an in-person consensus meeting was held with 25 panelists to decide on the essential items for reporting trial outcomes. The CONSORT Statement was followed by many other standards for reporting studies using a range of methodologies. To draw all these efforts together, the Enhancing the QUAlity and Transparency Of health Research (EQUATOR) Network was established in 2008 (Altman & Simera, 2016). The Network is an organization that brings together various parties with an interest in improving the quality of research publications. The EQUATOR Network has published guidance for developers of health research reporting guidelines (Moher et al., 2010). These recommend a review of the literature, a Delphi expert consensus study and a face-to-face consensus meeting.

consensus summarized in Table 6.1. These guidelines fulfilled a need for standardization of reporting to improve its quality. Formal consensus processes (Delphi method and consensus conference) were used to develop the guidelines. The level of innovation was not high. To achieve the aim of implementing the guidelines, a new global organizational structure (EQUATOR) had to be set up and sustained.

Case Example 6.5 involves the growing trend in a number of disciplines to set up registers of studies which allow methods and planned analyses to be laid out in advance of the study being done. Such preregistration aims to prevent researchers from presenting any findings after the data has been analysed as though it was predicted in advance.

Case Example 6.5 Preregistration of Research Studies

Recent decades have seen an increasing concern that many research findings cannot be replicated. This concern has been most strongly expressed in the social, behavioural and medical sciences. Many studies have been found not to replicate, leading scientists to ask why this is the case. These failures of replication have been attributed to common poor research practices, such as carrying out many exploratory statistical tests and focusing on any statistically significant results, not reporting findings that are contrary to the researcher's hypotheses, and stopping data collection as soon as statistical significance is achieved (Logg & Dorison, 2021). Some scientists have argued that the major problem is a failure to distinguish between prediction and postdiction when reporting results: "postdiction is characterized by the use of data to generate hypotheses about why something occurred, and prediction is characterized by the acquisition of data to test ideas about what will occur" (Nosek et al., 2018, p. 2600). Researchers may find their results more publishable if they present their post-hoc explanations (postdictions) as if they are predictions. However, replicability will be reduced because statistical significance testing is designed to test predictions rather than postdictions and will produce a high rate of false positive decisions if used post-hoc. A solution to this problem is for researchers to preregister an analysis plan prior to data collection, so that prediction and postdiction can be distinguished (Nosek et al., 2018). Registers have been set up in a range of disciplines to facilitate this, but these remain underused. Many medical journals now require preregistration of clinical trials as a condition of publication, which has increased the practice. Deliberative consensus has been used to determine the detailed procedures in preregistration. For example, a number of national psychological societies

(continued)

> **Case Example 6.5 (continued)**
> joined together to create a consensus template for the preregistration of quantitative research in psychology (Bosnjak et al., 2022). This template specifies the information that should be registered about a study's methods and the analysis plan. To develop the template, a task force from the psychological societies examined existing templates and had consensus-based discussions to select draft items. The draft was circulated for comment to a wide range of stakeholders and feedback was incorporated after in-depth discussion.

This example illustrates a number of characteristics of deliberative consensus on methods. The methodological problem was that no mechanism existed to allow preregistration and there was no agreement on what information about a study should be included on a register. This required coordination of efforts across multiple scientific organizations and the creation and maintenance of registration websites. Formal consensus processes have been used to determine what information about studies should be registered.

There are many other examples that could be given of the use of deliberative consensus processes to develop research methods. A number of these were described previously in Chap. 2, including the following:

- Case Example 2.6: A Protocol for Measuring the Volume of the Hippocampus with Magnetic Resonance Scans
- Case Example 2.7: International Organization for Standardization (ISO) Standards for Physical Measurements and Laboratory Equipment
- Case Example 2.8: Working Groups of the International Astronomical Union on Nomenclature and Measurement
- Case Example 2.17: Assessing the quality of systematic reviews using the AMSTAR checklist

Having now completed the description of the various ways that consensus processes have been used in science, the next chapter examines how an expert should be specified for the purposes of establishing consensus and what level of agreement should be regarded as consensus.

References

Altman, D. G., & Simera, I. (2016). A history of the evolution of guidelines for reporting medical research: The long road to the EQUATOR Network. *Journal of the Royal Society of Medicine, 109*(2), 67–77. https://doi.org/10.1177/0141076815625599

Baron, R. M., & Kenny, D. A. (1986). The moderator-mediator variable distinction in social psychological research: Conceptual, strategic, and statistical considerations. *Journal of Personality and Social Psychology, 51*(6), 1173–1182. https://doi.org/10.1037/0022-3514.51.6.1173

Bosnjak, M., Fiebach, C. J., Mellor, D., Mueller, S., O'Connor, D. B., Oswald, F. L., & Sokol, R. I. (2022). A template for preregistration of quantitative research in psychology: Report of the joint psychological societies preregistration task force. *American Psychologist, 77*(4), 602–615. https://doi.org/10.1037/amp0000879

Braun, V., & Clarke, V. (2006). Using thematic analysis in psychology. *Qualitative Research in Psychology, 3*, 77–101.

Braun, V., & Clarke, V. (2019). Reflecting on reflexive thematic analysis. *Qualitative Research in Sport, Exercise and Health, 11*(4), 589–597. https://doi.org/10.1080/2159676X.2019.1628806

Butcher, N. J., Monsour, A., Mew, E. J., Chan, A. W., Moher, D., Mayo-Wilson, E., et al. (2022). Guidelines for reporting outcomes in trial reports: The CONSORT-outcomes 2022 extension. *JAMA, 328*(22), 2252–2264. https://doi.org/10.1001/jama.2022.21022

Chavalarias, D., Wallach, J. D., Li, A. H., & Ioannidis, J. P. (2016). Evolution of reporting p values in the biomedical literature, 1990–2015. *JAMA, 315*(11), 1141–1148. https://doi.org/10.1001/jama.2016.1952

Cristea, I. A., & Ioannidis, J. P. A. (2018). P values in display items are ubiquitous and almost invariably significant: A survey of top science journals. *PLoS One, 13*(5), e0197440. https://doi.org/10.1371/journal.pone.0197440

Fisher, R. A. (1925). *Statistical methods for research workers.* Oliver and Boyd.

Fisher, R. A., & Yates, F. (1938). *Statistical tables for biological, agricultural and medical research.* Oliver and Boyd.

Hill, A. B. (1965). The environment and disease: Association or causation? *Proceedings of the Royal Society of Medicine, 58*(5), 295–300. https://doi.org/10.1177/003591576505800503

Kennedy-Shaffer, L. (2019). Before p < 0.05 to beyond p < 0.05: Using history to contextualize p-values and significance testing. *American Statistician, 73*(Suppl 1), 82–90. https://doi.org/10.1080/00031305.2018.1537891

Libby, W. F. (1946). Atmospheric helium three and radiocarbon from cosmic radiation. *Physical Review, 69*, 671–672. https://doi.org/10.1103/PhysRev.69.671.2

Logg, J. M., & Dorison, C. A. (2021). Pre-registration: Weighing costs and benefits for researchers. *Organizational Behavior and Human Decision Processes, 167*, 18–27. https://doi.org/10.1016/j.obhdp.2021.05.006

Moher, D., Schulz, K. F., Simera, I., & Altman, D. G. (2010). Guidance for developers of health research reporting guidelines. *PLoS Medicine, 7*(2), e1000217. https://doi.org/10.1371/journal.pmed.1000217

Mullis, K., Faloona, F., Scharf, S., Saiki, R., Horn, G., & Erlich, H. (1986). Specific enzymatic amplification of DNA in vitro: The polymerase chain reaction. *Cold Spring Harbour Symposia on Quantitative Biology, 51*(Pt 1), 263–273. https://doi.org/10.1101/sqb.1986.051.01.032

Nosek, B. A., Ebersole, C. R., DeHaven, A. C., & Mellor, D. T. (2018). The preregistration revolution. *Proceedings of the National Academy of Sciences U S A, 115*(11), 2600–2606. https://doi.org/10.1073/pnas.1708274114

Paxinos, G., & Watson, C. (1982). *The rat brain in stereotaxic coordinates*. Elsevier.

Turner, L., Shamseer, L., Altman, D. G., Schulz, K. F., & Moher, D. (2012). Does use of the CONSORT Statement impact the completeness of reporting of randomised controlled trials published in medical journals? A Cochrane review. *Systematic Reviews, 1*, 60. https://doi.org/10.1186/2046-4053-1-60

7

Specifying "Experts" and "Consensus"

A reader might well think that a book on "Expert Consensus in Science" should begin by explaining what is meant by "experts" and "consensus", rather than waiting halfway through to discuss these concepts in detail. Chapter 1 did give a brief definition of "expert consensus in science" as "a high level of agreement among scientists with relevant expertise about a specific scientific claim or about science-based practice or policy", which was sufficient to introduce the scope of this book. However, it raises additional questions about who is an "expert", and what level of agreement should be regarded as "consensus". The reason for leaving these issues until now is that some background knowledge is first needed on the range of uses that expert consensus has had in science and the various types of expertise that are required for each purpose.

7.1 Philosophers' Views on Who Is an Expert

A number of philosophers of science have considered the problem of determining who is an expert. A seminal work on the subject, by Alvin Goldman (2001), argued that experts in a particular domain "have more

© The Author(s) 2025
A. Jorm, *Expert Consensus in Science*, https://doi.org/10.1007/978-981-97-9222-1_7

beliefs… in true propositions and/or fewer beliefs in false propositions within that domain than most people do" (p. 91). To put it more simply, they have more accurate beliefs than most other people about the topic. However, having more beliefs in truths than falsities is not enough. Goldman also argued that there is a certain threshold of knowledge that must be reached for a person to be an expert. In other words, they must also have a substantial amount of knowledge about the topic. Goldman acknowledged that it is difficult to specify what this threshold is.

Other philosophers have disputed Goldman's view of expertise. Croce (2019) has summarized a number of these criticisms and proposed an alternative definition of who is an expert. According to his "research-oriented account", a person is an expert in a domain if they have "the capacity to contribute to the epistemic progress" in that domain and "can provide such help by offering true answers to the questions under dispute" in the domain (Croce, 2019, p. 4). In other words, expertise is not only a matter of knowing a lot about a topic, but also of being able to contribute to progress in the area through abilities in research. Under Goldman's (2001) definition, a teacher with extensive knowledge of an area of science could be an expert, but under Croce's (2019) definition they would need some capacity to create new knowledge through research as well.

The problem of determining who is an expert, and whether one person has greater expertise than another, differs depending on whether the person making the decision is a novice (e.g. layperson) or themselves an expert in the area. Goldman (2001, 2021) has discussed how a novice can judge these matters, proposing five possible sources of evidence that they could use:

1. Performance in a debate. When two possible experts disagree, a novice can try to identify which one has superior performance in a debate.
2. Agreement with other experts. A novice can check whether a possible expert comes to the same conclusions as other possible experts.
3. Appraisals by other experts. An expert will have been appraised by other experts through credentials such as specialized training from recognized institutions or professional registration.

4. Possible role of self-interest. An expert who has a strong self-interest in a particular position (e.g. they will profit from it) should be regarded as less trustworthy than one who does not.
5. Evidence of the expert's past track record. A novice can assess how successful a possible expert has been in the past in providing sound opinions.

Goldman concluded that the last of these sources of evidence (past track record) is the most important indicator of expertise for the layperson. Although this can be difficult for a layperson to judge, Goldman argued that there may be instances where this is possible. An example Goldman (2021) gives is where an astronomer predicted that a solar eclipse would be visible from a particular location at a specified time, and this was found to be true.

Goldman's (2001, 2021) proposed sources of evidence could also be used by other experts, who would be in a much better position to make these judgements than novices. However, other philosophers have discussed specifically how experts recognize expertise in their peers. Kitcher (1995) has distinguished two methods by which scientists attribute "authority" to peers. He calls these "direct calibration" and "indirect calibration". With direct calibration, a scientist uses their beliefs about the area of knowledge to evaluate the expertise of another person. In other words, the scientist concludes that the other person has expertise because their beliefs coincide with the scientist's own. By contrast, with indirect calibration, the scientist relies on the beliefs of other scientists, whose beliefs they have previously evaluated directly, about the other person. In this case, a scientist accepts the other person as an expert because other scientists they regard as authorities accept the person as such.

7.2 How Experts Have Been Selected in Consensus Studies

Although these philosophers have discussed what defines an expert and how this can be evaluated, their work has had remarkably little influence on how experts have been chosen for deliberative consensus studies, suggesting that these definitions are difficult to implement in practice. Rather, consensus studies have generally taken a pragmatic approach that relies on indirect indicators of having the relevant knowledge and skills. These indirect indicators are attributes that scientific experts typically have, such as professional training in a relevant discipline, employment by an organization involved in scientific research, membership of professional organizations in the area, being known to and communicating with other experts, being invited to speak at specialist conferences, publishing peer-reviewed papers, holding competitive grants and assessing other scientists' grants and papers. In some cases, scientific experts may receive accolades, such as fellowships, awards and prizes acknowledging a high level of contribution. Where experts are drawn from a science-based profession (e.g. a health practitioner), their expertise may involve skills (knowing-how, practical knowledge), as well as propositional knowledge (knowing-that, theoretical knowledge). This type of expertise may be indicated by professional training and registration, area of employment, years of practical experience and membership of a relevant professional organization. Many of these indicators have been used in the selection of experts, as illustrated by the examples in Table 7.1.

While the examples in Table 7.1 illustrate the use of a single indicator of expertise, it is more common for expert consensus studies to use multiple indicators. Case Example 7.1 illustrates this with a consensus study that involved a number of selection attributes.

Case Example 7.1 Use of Multiple Attributes to Select Experts on Performance Tests for the Exercise and Sport Sciences

Robertson et al. (2017) carried out a consensus study on the measurement properties and feasibility of performance tests for exercise and sport sciences. Experts were recruited from one of three categories: clinical exercise scientists/physiologists, sport scientists and academics. The clinical exercise scientists/physiologists had to have current accreditation with an accrediting body and relevant work experience in the industry. The sport scientists had to be currently employed by a professional sporting club or institution. The academics had to have publications on the topic. Taken together, these requirements illustrate a number of attributes that indicate expertise: professional qualifications, relevant work experience, accreditation and relevant publications.

Table 7.1 Examples of attributes indicating expertise in consensus studies

Attribute indicating expertise	Example consensus study's goal	How experts selected for example study
Professional qualifications, employment in field and work experience	Identify factors associated with high-level endurance performance (Konopka et al., 2022)	Professional athletes or coaches, exercise-scientists, physiotherapists, exercise-psychologists or medical physicians with extensive experience and knowledge of elite endurance performance
Membership of a scientific or professional organization	How to deal with air leak and intraoperative bleeding during thoracic surgery (Cardillo et al., 2022)	Members of the Italian Society of Thoracic Surgeons with known interest and high skills in thoracic surgery
Peer-reviewed publications	Ascertain scientists' views about attribution of global warming (Verheggen et al., 2014)	Publication of peer-reviewed or grey literature on global warming and climate change
Invited participation in a specialized symposium	Current status and future directions of mild cognitive impairment with regard to clinical presentation, cognitive and functional assessment, and the role of neuroimaging, biomarkers and genetics (Winblad et al., 2004)	Invitees to The First Key Symposium held in 2003, supported by The Royal Swedish Academy of Sciences and the *Journal of Internal Medicine*
Nomination by other experts	Recommend specific actions to end the COVID-19 public health threat (Lazarus et al., 2022)	Identification by project leaders of a core group of academic, health, NGO, government and policy experts from 25 countries; the core group then identified other individuals with expertise in COVID-19 from a diversity of countries

One factor that should not be used to select experts is their prior agreement with a particular conclusion, as this results in a "manufactured consensus". When this approach is used, it is better seen as signing a petition drafted by someone else rather than as a legitimate consensus exercise where alternative positions are given due consideration. Case Example 7.2 describes such a petition process with the World Climate Declaration.

Some of the attributes in Table 7.1 arguably indicate a higher level of expertise than others. As a general rule, publishing multiple papers on a topic is stronger than publishing a single paper, having highly cited publications is stronger than being less cited, being an office-holder or being in a higher grade of membership of a scientific organization is stronger than being an ordinary member, and personally carrying out relevant research is stronger than being a reader of the research. Sometimes it is possible to investigate consensus as a function of level of expertise. If the level of consensus increases with level of expertise, this arguably supports the consensus conclusions. As an illustration, Case Example 7.3 describes how the views of scientists on climate change were found to be associated with level of expertise as assessed by number of relevant publications and citations.

Case Example 7.2 Selection of Signatories to the World Climate Declaration

The World Climate Declaration, titled *There is No Climate Emergency*, is a statement supported by over 1600 scientists that disputes the conclusions of the Intergovernmental Panel on Climate Change (IPCC) (Climate Intelligence, 2023). Following are the conclusions of the declaration: natural as well as anthropogenic factors cause warming; warming is slower than predicted; climate policy relies on inadequate models; CO_2 is plant food, the basis of all life on Earth; global warming has not increased natural disasters; and climate policy must respect scientific and economic realities. The signatories include some eminent persons, including two Nobel Laureates. However, a check of the qualifications of the 75 Australian signatories found that most had no academic position or peer-reviewed research and were working in fields unrelated to climate science or the environment (RMIT ABC Fact Check, 2020). Applications to sign can be made through the World Climate Declaration website, and there is no avenue to express disagreement with specific conclusions.

Case Example 7.3 Level of Expertise and Views on Climate Change

A number of surveys have been carried out with climate scientists to assess the level of agreement on the role of human activity in climate change. These surveys show a very high level of agreement, but there is a dissenting minority. Anderegg et al. (2010) sought to assess whether scientists who have publicly endorsed the conclusions of the IPCC differ in level of expertise from scientists who have signed statements dissenting from the IPCC. The indicators of expertise they chose were publications on the topic and citations. The authors found that the scientists who supported the IPCC's conclusions were much more likely to have a high number of climate publications and a high number of citations to their top papers. For example, more than 90% of the IPCC-supporting scientists had 20 or more climate publications compared to around 20% for the dissenting group.

Looking at the various attributes in Table 7.1, one might ask whether there is any common defining feature for who is an expert. One factor behind most of the attributes is acknowledgement of expertise by peers. Such acknowledgement is involved in acquiring scientific and professional credentials, admission to professional societies, authorship of publications that have been peer-reviewed, invited participation in a specialized symposium, and nomination by other experts. A general definition of scientific expert might therefore be "a person who is acknowledged as reaching a standard of expertise on a scientific topic by the community of scientists working on that topic". Such a definition would be consistent with Kitcher's (1995) proposal for how scientists can use direct or indirect calibration to evaluate whether a person is an expert on a topic. It is also compatible with a number of the sources of evidence that Goldman (2001, 2021) proposed that novices could use, including agreement with other experts, appraisal by other experts and past track record.

When proposing the centrality of peer acknowledgement to specifying who is an expert, we must also note that sometimes people with relevant expertise have not been acknowledged for this because of biases in the community; for example, in the past women were not always acknowledged as experts (Fricker, 2007). The importance of selecting a diverse group of experts for coming to optimal conclusions is a topic that is covered in more detail in Chap. 9.

7.3   Provision of Additional Data to Strengthen Expertise

When scientific experts are selected for a consensus study, they are chosen for the expertise they bring to the process. However, in some cases they are also provided with additional data relevant to the topic of the consensus, which they can draw on if they wish. Such data might include systematic reviews of the literature, for example on randomized controlled trials of an intervention, or a compilation of specific data values from existing studies where a quantitative estimate is required. To critically interpret such data itself requires a high level of relevant expertise. To illustrate how additional data can be used to strengthen expertise, Case Example 7.4 describes a consensus process where a narrative literature review was provided to experts, while Case Example 7.5 describes one where data values from a systematic review of the literature were provided to inform experts' quantitative estimates.

Case Example 7.4 Use of a Literature Review to Inform Experts on Campaigns to Reduce Mental Health-Related Stigma

Clement et al. (2010) carried out a consensus study to find messages suitable for use in campaigns to reduce stigma towards people with a mental illness. The experts were delegates attending an International Stigma Conference who all had expertise in mental health-related stigma. Prior to the conference, the research team reviewed the academic and campaign literature to find out the main types of messages used. They found ten types of messages, such as "valuing difference" messages, "biomedical" messages, "recovery" messages and "social inclusion" messages. The team also prepared an overview of the research evidence, which covered different types of evidence: intervention studies, association studies, qualitative studies and studies of opinion. The main conclusion from the review was that "biomedical" messages reduce blame but are problematic in other ways, and that "recovery-oriented" messages appear promising. When the experts met, the ten types of messages were explained by the researchers and then the experts used electronic voting pads to rate each message type for whether it should be used. Graphical feedback was immediately provided on the outcomes of the voting. Next, the review of research evidence was presented to the experts in the form of a lecture and there was a facilitated discussion about the messages where there was least consensus. In a final round of voting, the most highly rated message types were "recovery-oriented" and "see the person" messages.

Case Example 7.5 Use of Published Data to Inform Experts on Estimating the Global Prevalence of Dementia

Ferri et al. (2005) carried out an expert consensus study to estimate the prevalence of dementia for each world region. Twelve international experts were recruited and provided with a systematic review of published studies on the prevalence of dementia. The review gave the age-specific prevalence rates from each study along with methodological details such as the sample size, the sampling procedure, the response rate and the diagnostic criteria used. The experts were asked to review the document and give their own estimates for prevalence in each region in five-year age bands from 60 to 64 years up to 85 years and older. Because the data were sparse or non-existent for some regions, the experts were requested to make inferences using the data from other regions and to provide comments on the decisions they made. The group response was summarized as the mean prevalence estimate and returned to the experts along with the comments provided by others. The experts were given the opportunity to revise their estimates if they wished. The second-round estimates were used to provide the final consensus prevalence rates and a global estimate of the number of dementia cases.

The provision of additional data for consideration by expert panels has implications for the usefulness of hierarchies of evidence that place randomized controlled trials at the top and expert consensus at the bottom (see Chap. 6). These hierarchies do not take account of the types of data that the experts are drawing on, which can range from a systematic review of randomized controlled trials through to their own professional experience. Similarly, it calls into question the distinction in clinical practice guidelines between "evidence-based guidelines" and "consensus-based guidelines" (see Chap. 6). Even where randomized controlled trial data exists, it does not directly determine professional practice and has to be interpreted by a panel of experts for its relevance to typical patients (who often have co-morbid diseases), to people of different ethnicities and to the health system of the country.

7.4   Specifying Experts on Values

A number of philosophers of science have pointed out that science is not value free (Douglas, 2009; Kitcher, 2011; Longino, 2004). This can be seen, for example, in decisions about what topics are important to prioritize in research. Chapter 6 argued that values become particularly important in areas of professional practice and policy, where deciding a course of action involves a consensus on values in addition to a consensus on scientific facts. Scientists and science-based professionals are in a position to make judgements about both the scientific evidence and the values involved in deciding what actions should be taken. However, in many cases, the values of these experts may differ from those of the people who are potentially affected by any consensus on what should be done.

Surveys of scientists in a number of countries show that they differ in attitudes from the general public, being more likely to have liberal or left-wing political attitudes (van de Werfhorst, 2020) and being less religious (Ecklund et al., 2016; Masci, 2009; Nakhaie & Brym, 1999). These differences are greater in some disciplines than others, with liberal attitudes particularly common in the social sciences (Klein & Stern, 2005; Nakhaie & Brym, 1999). It has been argued that greater political diversity would be beneficial to social psychological science, as it would reduce bias in selection of topics and in conclusions, and it would empower dissenting minority views and improve critical thinking (Duarte et al., 2015).

Scientists and science-informed professionals also tend to come from certain cultural groups, particularly high-income Western countries. As a consequence, their values may not reflect those of people in other cultures or of cultural minorities in their own societies. Similarly, in areas of professional practice like medicine, the values of professionals may differ from those of patients and family members who are directly affected by the health condition in question.

Given the importance of getting a consensus on the values of all interested parties, some consensus studies include the general public, cultural experts and consumer advocates as experts in matters of values.

Case Example 7.6 describes how surveys of the public in a wide range of countries were used to develop disability weights for various health conditions, which could be used to estimate the global burden of diseases. This work was undertaken to replace earlier disability weights produced by health professionals, but has itself been controversial because of doubts about the validity of the weights for some health conditions.

Case Example 7.6 Use of Laypeople to Develop Disability Weights for the Global Burden of Disease Study

The Global Burden of Disease (GBD) study aims to investigate the impact of different diseases and injuries on population health, producing estimates for all the countries and regions of the world, as well as global estimates. Central to the GBD study is the concept of the disability-adjusted life year (DALY), which is a way of quantifying losses of healthy life due to either premature mortality or time lived in a state of reduced functioning (disability). The study estimates the DALYs due to various diseases in order to quantify their relative impact on a population. To estimate DALYs, the researchers use "disability weights", where each health condition is rated on a scale from 0 (no loss of health) to 1 (loss equivalent to death). The GBD study began in the early 1990s and has been regularly updated since. In the early GBD work, the disability weights were produced by a panel of health professionals in a deliberative consensus exercise. However, this approach was criticized as failing to reflect the perspectives of diverse cultures and environments, and eventually replaced by a new study of disability weights produced by the ratings of members of the public from many countries (Salomon, 2010). To produce the disability weights, the lay participants had to compare pairs of hypothetical individuals with different health conditions and say which one they thought was healthier. Because the participants were not health professionals, they had no expertise on the impact of various health conditions and had to judge this from brief lay descriptions provided by the researchers. This led to some anomalies, for example for complete hearing loss and neck-level spinal cord lesions, where the weights were unrealistically small given the known disability from measures taken on patients. Subsequent research had to be carried out with revised descriptions of these health conditions to produce more realistic weights (Salomon et al., 2015).

Another related body of research involves the use of quality-adjusted life years (QALYs) to estimate the impact of various health conditions on quality of life. QALYs are widely used in health economics where the impact of health interventions, such as pharmaceuticals, is costed in terms of dollars per QALY gained from the intervention. However, judgements about quality of life may be influenced by culture, as illustrated in Case Example 7.7. Such cultural differences imply that for some values, judgements may not be culturally transportable.

Another group which has expertise to offer in matters of values is consumer advocates. The importance of including consumer perspectives is becoming increasingly recognized in the area of health. Health interventions need to be acceptable to the people they are targeted at, and these consumers may have a different values perspective to health professionals. One way to ensure that a consensus is consistent with their values is to require a consensus of both professional and consumer experts about actions that should be taken. Case Example 7.8 illustrates the use of consumer advocates as experts along with professionals in the area of mental health first aid for a suicidal person.

Case Example 7.7 Cultural Differences in the Valuation of Health States

To measure a QALY requires that values be assigned to states of health, which can vary on a scale between 1 for perfect health and 0 for dead. A person who lives in perfect health for 1 year will have 1 QALY, as will a person who lives with half perfect health for 2 years. To develop the values of various health states, members of the public are given descriptions of health involving how affected the person is in areas such as physical functioning, role limitation, social functioning, pain, mental health and vitality. Various methods are used to convert these values to the 0 to 1 scale. Studies to develop these values have been carried out in a range of countries. A review of these studies (Wang & Poder, 2023) found that the valuations of health states varied across countries. Anglo-Saxon countries were found to give more importance to pain when rating quality of life, while non-Anglo-Saxon countries gave more importance to physical functioning. Furthermore, countries with a higher level of economic development cared less about physical functioning but more about mental health and pain.

Case Example 7.8 Consumer Advocates as Experts on Mental Health First Aid for a Suicidal Person

If a person becomes suicidal, their friends and family are in the best position to provide initial assistance, before professional help can be accessed. To guide the content of training in suicide first aid for the public, an expert consensus study was carried out on what actions are likely to be helpful (Ross et al., 2014). A systematic search was carried out of the suicide prevention literature to find statements about possible helping actions. In the research, 436 statements about actions were found, and experts were asked to rate these for their importance. Two expert panels were recruited. The first involved professionals working in the area of suicide prevention, and the second consisted of consumers who had personal experience of being suicidal and had taken on an advocacy role. Consensus on a helping statement required that it had to be endorsed as important or essential by at least 80% of both the professional and consumer groups. A total of 164 statements met this criterion and have been used internationally to guide the content of Mental Health First Aid training.

A potential stumbling block in involving non-scientists in value-based consensus processes is that these require an understanding of the scientific evidence as a foundation when making the value judgements. For many issues, the public will lack the required knowledge. One approach to overcoming this limitation is the use of citizen juries. A citizen jury involves a randomly chosen or diverse group of citizens who are given a specific policy issue to make recommendations on. The policy issue can be a science-based one. The jury is given a detailed briefing by experts about the issue and can ask them questions. They are asked to deliberate on the issue and to produce recommendations. If the jury is randomly chosen, its recommendations are more likely to be acceptable to the wider public. Case Example 7.9 describes the use of citizen juries on the controversial area of onshore wind farms in Scotland. A strength of this example is that it involved multiple juries in different communities, allowing an assessment of the reliability of recommendations across communities.

Case Example 7.9 Citizen Juries on Onshore Wind Farms in Scotland

Engagement of the public is emphasized in Scottish Planning Policy. Citizen juries have been trialed as a means of achieving this for the controversial issue of wind farms (Roberts & Escobar, 2015). Juries were formed in three different communities which had varying degrees of exposure to wind farm developments. The jurors were selected using quotas to ensure that they were broadly representative of the Scottish population in sociodemographic characteristics and attitudes towards the environment. The juries were set the following task: "There are strong views on wind farms in Scotland, with some people being strongly opposed, others being strongly in favour and a range of opinions in between. What should be the key principles for deciding about wind farm development, and why?" The juries met for two days separated by a "reflection phase" of two to three weeks. On the first day, they heard from an impartial expert on energy and climate change, followed by two experts presenting arguments for or against wind power, and then two experts presenting arguments for or against wind farms. On the second day, two to three weeks later, the jury had to set the agenda and work together on the assigned task of coming up with the key principles. Many jury members reported changing their views during the process and the three juries were found to produce many similar principles. The principles included the desirable energy mix for Scotland, the characteristics of evidence needed for decision-making, the range of positive and negative impacts that should be taken into account and the public responsibility for reducing energy consumption.

7.5   Specifying What Is Consensus

Chapter 4 distinguished between spontaneous and deliberative consensus. With spontaneous consensus, there is no formal process to ascertain agreement, but there are indicators that it has occurred, such as a high rate of positive citations and incorporation in textbooks. In the case of citations, there is evidence that they reflect the importance of a scientific publication. This can be seen in the area of astronomy, where the American Astronomical Society marked its centennial by asking 53 senior astronomers to nominate the most important articles in the past century from two major journals (Abt, 2000). The citations to these articles were compared to those of adjacent articles published at the same time, which served as controls. The nominated articles averaged 6.7 times as many citations as the control articles and had citation half-lives that averaged 2.5 times longer.

Earlier chapters presented examples of spontaneous consensus around findings that had a very high rate of citation and were awarded Nobel Prizes (Case Example 4.1 on *Accelerating Expansion of the Universe* and Case Example 4.2 on *Bacterial Infection and Stomach Ulcers*) and methodological contributions in textbooks that have become very widely used (Case Example 6.1 on *The P < .05 Level of Statistical Significance* and Case Example 6.2 on *Atlas of the Rat Brain*).

Where consensus is formally assessed in a deliberative process, it can range from 0% to 100% of experts agreeing with a claim. A value closer to 100% is a stronger endorsement, but it has been argued by Beatty and Moore (2010) that unanimity is not necessarily a good thing:

> The existence of a persistent minority indicates that the majority position was tested, which is confidence-inspiring; and that the minority position was bested, but not to the point that its advocates capitulated on the substantive issue. A mixed outcome is a record and a reminder that there are alternative points of view to the one that prevailed at the time of the vote, and serves as a way to keep an issue alive. (p. 203)

According to this view, the existence of dissent signals that there has been full consideration of alternative positions and that the issues have been debated. Similarly, Dellsén (2021) has argued that the existence of some dissent shows that there has not been pressure from others in the group to conform:

> Generally speaking, however, a certain marginal level of dissent among experts should increase a layperson's confidence in a theory about which the experts have otherwise reached a consensus, roughly because this indicates that the agreement on the theory in question is less likely to be due to a conformity effect, that is, 'groupthink'. (p. 20)

A push for unanimity can also lead scientists to a bland consensus, as that is what everyone can agree on. This is what Sarewitz (2011) experienced when working on a consensus report on *Geoengineering: A National Strategic Plan for Research on Climate Remediation*. In a reflection on the consensus process, he concluded:

The very idea that science best expresses its authority through consensus statements is at odds with a vibrant scientific enterprise. Consensus is for textbooks; real science depends for its progress on continual challenges to the current state of always-imperfect knowledge. Science would provide better value to politics if it articulated the broadest set of plausible interpretations, options and perspectives, imagined by the best experts, rather than forcing convergence to an allegedly unified voice.... Unlike a pallid consensus, a vigorous disagreement between experts would provide decision-makers with well-reasoned alternatives that inform and enrich discussions as a controversy evolves, keeping ideas in play and options open. (p. 7)

However, unanimity is not incompatible with a full consideration of alternatives and lack of pressure to come to a consensus. An indicator of this occurring would be where multiple claims about a particular topic are judged by the experts, and only some reach unanimity. If some claims are very strongly endorsed and others not, this indicates that dissent on issues was acceptable to the group. Case Example 7.10 describes an expert consensus study on what parents should do to prevent body dissatisfaction in their young children. The level of consensus varied greatly depending on the statement being rated, with some achieving unanimous endorsement and others not even managing majority support.

Case Example 7.10 Varying Levels of Consensus on What Parents Should Do to Prevent Body Dissatisfaction in Preschool Children

Hart et al. (2014) carried out an expert consensus study on what parents of preschool children should do to prevent body dissatisfaction and unhealthy eating patterns. A group of 28 experts rated 335 potential preventive actions for parents that were found from a search of the literature. Statements about prevention had to be endorsed by at least 90% of the panel to be included as reaching consensus. The levels of endorsement as "essential" or "important" varied widely. Some statements were unanimously endorsed: for example, "Parents should accept their child as they are, regardless of the child's weight, size, or body shape" and "Parents should discourage their child from dieting as a way of losing weight". However, others were rejected by a majority of the experts: for example, "Parents should use a goal chart to encourage exercise habits in family members, and make sure there are fun rewards" and "Parents should promote to their child that physical activity is a means of losing weight". This wide variation indicates that unanimity was not achieved through group pressure or exclusion of alternative views.

Where a cut-off for percentage agreement is used to define consensus, the cut-off chosen may depend on the question asked and the purpose of the consensus exercise. Chapter 3 described the work of Vickers (2023), which aims to identify scientific facts that will be forever true (although perhaps subject to minor fine-tuning). His proposed criteria for identifying these are "a solid scientific consensus amounting to at least 95%, in a scientific community that is large, international, and diverse" (p. 18), which is a very high bar. Consensus on methodological standards is another area where a very high level of consensus would be appropriate.

For other issues, such as defining a concept or measure, a simple majority might be appropriate. In these cases, the role of a consensus exercise is not about "truth" but about a scientific community agreeing that unless they agree on something in their field, it cannot advance efficiently. In such instances, if all the experts who are voting agree to abide by the majority decision, even if not everybody loves it, and this decision is recognized by others in the field, a higher level of consensus may not be necessary. Case Example 7.11 illustrates how a majority vote was used by the International Astronomical Union to remove Pluto from the list of planets, while Case Example 7.12 describes how the International Bureau of Weights and Measures uses a majority vote to update the International System of Units (SI).

Case Example 7.11 Voting to Remove Pluto from the List of Planets

Pluto was removed from the list of planets by the International Astronomical Union (IAU) in 2006 (Zachar & Kendler, 2012). What prompted its reclassification was the discovery in 2005 of Eris, which is located beyond Pluto, is larger and has a moon, making it potentially a tenth planet. However, its orbital path is quite different to other planets, with a 45 degrees inclination relative to the orbital plane of the Earth. This led the IAU to reconsider the definition of a planet. An IAU Working Group on the Definition of a Planet was unable to come to a consensus on the issue. Eventually, it was put to a vote at the IAU's 2006 conference, with 78.5% of members who were present voting for a definition of a planet which required that it orbit the Sun (not be a moon), be massive enough to take a spherical shape, and not be a member of a larger group of objects sharing the same orbital location. The latter requirement was not met by Pluto and Eris, which were put into a new category of dwarf planets. The vote was controversial because it occurred on the last day of the conference when most delegates had already left and only 424 members voted.

Case Example 7.12 Voting on Changes to the International System of Units (SI)

The International Bureau of Weights and Measures is an intergovernmental organization in which member countries set worldwide measurement standards, including for the SI system of units which is the modern form of the metric system (OECD/BIPM, 2020). The Bureau defines the base units of the SI system, such as the second, metre, kilogram and ampere. The ultimate governing body for the Bureau is the General Conference on Weights and Measures, which consists of delegates from all the member countries and meets every four years. The Conference adopts resolutions based on an absolute majority (50%) of votes of member countries attending the meeting, with the resolutions then implemented by the member countries. The Bureau has voted to change the definitions of the SI units on a number of occasions throughout its history in order to give greater precision.

Another factor in determining what is an appropriate level of consensus is the degree of risk associated with a particular course of action. In areas of environmental and health policy, policymakers sometimes invoke the "precautionary principle", which "holds that we should not allow scientific uncertainty to prevent us from taking precautionary measures in response to potential threats that are irreversible and potentially disastrous" (Resnik, 2003, p. 329). Intemann (2017) has discussed the implications of taking account of risk assessment when using scientific consensus in policy decisions:

> it is also dangerous to think that scientific consensus is necessary for making rational policy decisions. Often, there are contexts in which we must adopt public policies despite significant uncertainties and thus despite a lack of consensus about the scientific evidence. For example, significant, and legitimate, disagreement might exist about the extent to which a particular substance is toxic. Nonetheless, one might still argue that in the face of uncertainty, certain risks are more acceptable than others. Hence, it might be rational to adopt regulations even if there is no scientific consensus about toxicity. (p. 199)

The most prominent contemporary example of the application of the precautionary principle is in the area of climate change. There are potentially catastrophic risks involved in not taking action to reduce global warming, even if there are scientific uncertainties. In this case, it is arguably unwise to require a very high level of scientific consensus before taking action. Conversely, some solutions involving climate engineering (e.g. ocean fertilization to remove CO_2 and solar radiation management to reflect sunlight back into outer space) are also potentially risky, and a very high level of certainty might be required before implementing them. Case Example 7.13 describes another issue, the health effects of electromagnetic fields, where the precautionary principle has been invoked by some experts.

Case Example 7.13 Application of the Precautionary Principle with Health Effects of Electromagnetic Fields

Modern technologies (e.g. mobile phones) have increasingly exposed humans to electromagnetic fields (EMF). The World Health Organization (WHO) has set up the International EMF Project to assess the health and environmental effects of EMF exposure. The WHO (2005) has advised that "No major public health risks have emerged from several decades of EMF research, but uncertainties remain" (p. 1). However, in 2009, a scientific panel independent of WHO met in Seletun, Norway, for three days to discuss the scientific evidence and the public health implications of global exposures to artificial EMF. This scientific panel published a consensus statement that incorporated the precautionary principle and made recommendations for action that went beyond the cautious conclusion of WHO (Fragopoulou et al., 2010, p. 2):

> The Seletun Scientific Panel ... recommends preventative and precautionary actions that are warranted now, given the existing evidence for potential global health risks. We recognize the duty of governments and their health agencies to educate and warn the public, to implement measures balanced in favor of the Precautionary Principle, to monitor compliance with directives promoting alternatives to wireless, and to fund research and policy development geared toward prevention of exposures and development of new public safety measures.

These contrasting conclusions illustrate how an assessment of risks may alter the scientific consensus on policy action.

To answer the question posed in this section about what level of agreement should define "consensus", the purpose of the consensus exercise needs to be considered. For some purposes (e.g. defining scientific truth or best practice in methodology), a very high threshold for consensus is appropriate, whereas for others (e.g. defining concepts or standardizing measures) a much lower threshold may be adequate. While a consensus approaching 100% indicates a strong level of endorsement by experts, it is important that open discussion and dissent are allowed and that there is no pressure to conform. A high level of consensus that is less than 100% is an indicator that these conditions are met and may be considered a strength in many circumstances. A lower level of consensus may also be appropriate in matters of public policy where a prompt decision needs to be made.

Levels of agreement lie on a continuum from <50% to 100%. However, it may be useful to divide this continuum into bands and consider what types of use they might be appropriate for. Table 7.2 proposes such a classification, which should be regarded as an interim proposal in the absence of any deliberative process to formalize it.

Table 7.2 A proposal for classification of levels of agreement among experts

Level of agreement	Description of level	Potential uses
100%	Unanimity	Defining truth (but requires opportunities for dissent to be credible)
95%+	High level of consensus	Defining truth
80%+	Consensus	Establishing the current thinking on likely truth or an acceptable methodology
67%+	Substantial majority agreement	Coming to a decision on important matters of social or health policy where the "precautionary principle" applies
50%+	Majority agreement	Defining terms and measures

7.6   Reporting on Consensus

In order to critically evaluate consensus studies, it is important that the way that experts have been selected, the criteria for consensus, and the process for arriving at it are fully reported. However, it is surprising that many studies that use a consensus methodology do not report these basic details, opening them to criticism that alternative positions were not fully considered or there was pressure to conform. Case Example 7.14 illustrates this with a position statement by the Sleep Research Society on the adoption of permanent standard time in the United States.

To improve the quality of reporting, the EQUATOR Network (see Chap. 6, Case Example 6.4) has recently published guidelines for reporting consensus studies in biomedicine (Gattrell et al., 2024). These were developed based on a Delphi expert consensus study. The guidelines involve 35 features of a consensus study which should be reported, including how expert panel members were selected (including criteria for panelist inclusion, how they were recruited, any roles as members of the public, patients or carers) and assessment of consensus (including how questions were presented, the definition of consensus, how responses

---

Case Example 7.14 Lack of Information on Level of Consensus in the Sleep Research Society's Position Statement on the Adoption of Permanent Standard Time in the United States

The Sleep Research Society is a US professional organization concerned with education and research on sleep and circadian science. The Society has published an official position statement in one of its journals on the issue of whether daylight saving time should be continued in the United States (Malow, 2022). The Position Statement briefly reviews the evidence on the health effects of changing the clock, including effects on sleep loss, well-being, stroke, myocardial infarction and traffic accidents. Because of the health problems associated with changes to daylight saving time, the Society has advocated for the adoption of permanent standard time. Although this is an official position of the Society, no information is provided about who produced the Statement and how it came to be adopted, including how experts were chosen, how consensus was established and the level of consensus achieved.

were collected and synthesized, how any feedback was presented to panel members, whether voting was anonymous).

The quality of methods to establish a scientific consensus is critical to deciding whether any conclusions should be accepted by scientific peers, practitioners, policymakers and the public. Clear reporting on what was done is the first step in evaluating quality. However, it is not sufficient, as some methods are more likely to result in quality decisions than others. The weaknesses of many consensus methods and how these might be improved are the subject of the next three chapters of this book.

References

Abt, H. A. (2000). The most frequently cited astronomical papers published during the past decade. *Bulletin of the American Astronomical Society, 32*, 937–941.

Anderegg, W. R., Prall, J. W., Harold, J., & Schneider, S. H. (2010). Expert credibility in climate change. *Proceedings of the National Academy of Sciences U S A, 107*(27), 12107–12109. https://doi.org/10.1073/pnas.1003187107

Beatty, J., & Moore, A. (2010). Should we aim for consensus? *Episteme, 7*(3), 198–214. https://doi.org/10.3366/epi.2010.0203

Cardillo, G., Nosotti, M., Scarci, M., Torre, M., Alloisio, M., Benvenuti, M. R., et al. (2022). Air leak and intraoperative bleeding in thoracic surgery: A Delphi consensus among the members of Italian society of thoracic surgery. *Journal of Thoracic Disease, 14*(10), 3842–3853. https://doi.org/10.21037/jtd-22-619

Clement, S., Jarrett, M., Henderson, C., & Thornicroft, G. (2010). Messages to use in population-level campaigns to reduce mental health-related stigma: Consensus development study. *Epidemiologia e Psichiatria Sociale, 19*(1), 72–79. https://doi.org/10.1017/s1121189x00001627

Climate Intelligence. (2023). *World climate declaration: There is no climate emergency*. CLINTEL. Retrieved February 21, 2023, from https://clintel.org/world-climate-declaration/

Croce, M. (2019). On what it takes to be an expert. *The Philosophical Quarterly, 69*(274), 1021. https://doi.org/10.1093/pq/pqy044

Dellsén, F. (2021). Consensus versus unanimity: Which carries more weight? *British Journal for the Philosophy of Science*. https://doi.org/10.1086/718273

Douglas, H. E. (2009). *Science, policy, and the value-free ideal.* University of Pittsburgh Press. Table of contents only http://www.loc.gov/catdir/toc/fy1002/2009005463.html

Duarte, J. L., Crawford, J. T., Stern, C., Haidt, J., Jussim, L., & Tetlock, P. E. (2015). Political diversity will improve social psychological science. *Behavioral and Brain Sciences, 38,* e130. https://doi.org/10.1017/S0140525X14000430

Ecklund, E. H., Johnson, D. R., Scheitle, C. P., Matthews, K. R. W., & Lewis, S. W. (2016). Religion among scientists in international context: A new study of scientists in eight regions. *Socius: Sociological Research for a Dynamic World, 2,* 1–9. https://doi.org/10.1177/2378023116664353

Ferri, C. P., Prince, M., Brayne, C., Brodaty, H., Fratiglioni, L., Ganguli, M., et al. (2005). Global prevalence of dementia: A Delphi consensus study. *Lancet, 366*(9503), 2112–2117. https://doi.org/10.1016/s0140-6736(05)67889-0

Fragopoulou, A., Grigoriev, Y., Johansson, O., Margaritis, L. H., Morgan, L., Richter, E., & Sage, C. (2010). Scientific panel on electromagnetic field health risks: Consensus points, recommendations, and rationales. *Reviews on Environmental Health, 25*(4), 307–317. https://www.ncbi.nlm.nih.gov/pubmed/21268443

Fricker, M. (2007). *Epistemic injustice: Power and the ethics of knowing.* Oxford University Press. Table of contents only http://www.loc.gov/catdir/toc/ecip0710/2007003067.html

Gattrell, W. T., Logullo, P., van Zuuren, E. J., Price, A., Hughes, E. L., Blazey, P., et al. (2024). ACCORD (ACcurate COnsensus Reporting Document): A reporting guideline for consensus methods in biomedicine developed via a modified Delphi. *PLoS Medicine, 21*(1), e1004326. https://doi.org/10.1371/journal.pmed.1004326

Goldman, A. I. (2001). Experts: Which ones should you trust? *Philosophy and Phenomenonological Research, 63,* 85–110. https://doi.org/10.1111/j.1933-1592.2001.tb00093.x

Goldman, A. I. (2021). How can you spot the experts? An essay in social epistemology. *Royal Institute of Philosophy Supplement, 89,* 85–98. https://doi.org/10.1017/S1358246121000060

Hart, L. M., Damiano, S. R., Chittleborough, P., Paxton, S. J., & Jorm, A. F. (2014). Parenting to prevent body dissatisfaction and unhealthy eating patterns in preschool children: A Delphi consensus study. *Body Image, 11*(4), 418–425. https://doi.org/10.1016/j.bodyim.2014.06.010

Intemann, K. (2017). Who needs consensus anyway? Addressing manufactured doubt and increasing public trust in climate science. *Public Affairs Quarterly, 31*(3), 189–208. https://doi.org/10.2307/4473279210.2307/44732792

Kitcher, P. (1995). *The advancement of science: Science without legend, objectivity without illusions.* Oxford University Press. https://doi.org/10.1093/0195096533.001.0001

Kitcher, P. (2011). *Science in a democratic society.* Prometheus Books.

Klein, D. B., & Stern, C. (2005). Professors and their politics: The policy views of social scientists. *Critical Review, 17*, 257–303. https://doi.org/10.1080/08913810508443640

Konopka, M. J., Zeegers, M. P., Solberg, P. A., Delhaije, L., Meeusen, R., Ruigrok, G., et al. (2022). Factors associated with high-level endurance performance: An expert consensus derived via the Delphi technique. *PLoS One, 17*(12), e0279492. https://doi.org/10.1371/journal.pone.0279492

Lazarus, J. V., Romero, D., Kopka, C. J., Karim, S. A., Abu-Raddad, L. J., Almeida, G., et al. (2022). A multinational Delphi consensus to end the COVID-19 public health threat. *Nature, 611*(7935), 332–345. https://doi.org/10.1038/s41586-022-05398-2

Longino, H. E. (2004). How values can be good for science. In P. K. M. G. Wolters (Ed.), *Science, values, and objectivity* (pp. 127–142). University of Pittsburgh Press.

Malow, B. A. (2022). It is time to abolish the clock change and adopt permanent standard time in the United States: A Sleep Research Society position statement. *Sleep, 45*(12), 1–4. https://doi.org/10.1093/sleep/zsac236

Masci, D. (2009). *Scientists and belief.* https://www.pewresearch.org/religion/2009/11/05/scientists-and-belief/#:~:text=Finally

Nakhaie, M. R., & Brym, R. J. (1999). The political attitudes of Canadian professors. *Canadian Journal of Sociology, 24*, 329–353. https://doi.org/10.2307/3341393

OECD/BIPM. (2020). *International regulatory co-operation and international organisations: The case of the International Bureau of Weights and Measures (BIPM).* https://www.oecd.org/en/topics/international-regulatory-co-operation.html

Resnik, D. B. (2003). Is the precautionary principle unscientific? *Studies in History and Philosophy of Biological and Biomedical Sciences, 34*, 329–344. https://doi.org/10.1016/S1369-8486(02)00074-2

RMIT ABC Fact Check. (2020, February 27). Who are the 75 Australian 'scientists and professionals' who say there is no climate emergency? *ABC News.*

https://www.abc.net.au/news/2020-02-27/who-are%2D%2Dscientists-professionals-who-say-no-climate-emergency/11734966

Roberts, J., & Escobar, O. (2015). *Involving communities in deliberation: A study of three citizens' juries on onshore wind farms in Scotland.* https://www.climatexchange.org.uk/wp-content/uploads/2023/09/citizens_juries_report_exec_summary.pdf

Robertson, S., Kremer, P., Aisbett, B., Tran, J., & Cerin, E. (2017). Consensus on measurement properties and feasibility of performance tests for the exercise and sport sciences: A Delphi study. *Sports Medicine - Open, 3*(1), 2. https://doi.org/10.1186/s40798-016-0071-y

Ross, A. M., Kelly, C. M., & Jorm, A. F. (2014). Re-development of mental health first aid guidelines for suicidal ideation and behaviour: A Delphi study. *BMC Psychiatry, 14,* 241. https://doi.org/10.1186/s12888-014-0241-8

Salomon, J. A. (2010). New disability weights for the global burden of disease. *Bulletin of the World Health Organization, 88*(12), 879. https://doi.org/10.2471/BLT.10.084301

Salomon, J. A., Haagsma, J. A., Davis, A., de Noordhout, C. M., Polinder, S., Havelaar, A. H., et al. (2015). Disability weights for the Global Burden of Disease 2013 study. *Lancet Global Health, 3*(11), e712–e723. https://doi.org/10.1016/S2214-109X(15)00069-8

Sarewitz, D. (2011). The voice of science: Let's agree to disagree. *Nature, 478*(7367), 7. https://doi.org/10.1038/478007a

van de Werfhorst, H. G. (2020). Are universities left-wing bastions? The political orientation of professors, professionals, and managers in Europe. *British Journal of Sociology, 71,* 47–73. https://doi.org/10.1111/1468-4446.12716

Verheggen, B., Strengers, B., Cook, J., van Dorland, R., Vringer, K., Peters, J., et al. (2014). Scientists' views about attribution of global warming. *Environmental Science & Technology, 48*(16), 8963–8971. https://doi.org/10.1021/es501998e

Vickers, P. (2023). *Identifying future-proof science.* Oxford University Press.

Wang, L., & Poder, T. G. (2023). A systematic review of SF-6D health state valuation studies. *Journal of Medical Economics, 26*(1), 584–593. https://doi.org/10.1080/13696998.2023.2195753

Winblad, B., Palmer, K., Kivipelto, M., Jelic, V., Fratiglioni, L., Wahlund, L. O., et al. (2004). Mild cognitive impairment – Beyond controversies, towards a consensus: Report of the International Working Group on Mild Cognitive Impairment. *Journal of Internal Medicine, 256*(3), 240–246. https://doi.org/10.1111/j.1365-2796.2004.01380.x

World Health Organization. (2005, May 23). *Electromagnetic fields project.* WHO. Retrieved 18 Aug 2024 from https://www.who.int/initiatives/the-international-emf-project

Zachar, P., & Kendler, K. S. (2012). The removal of Pluto from the class of planets and homosexuality from the class of psychiatric disorders: A comparison. *Philosophy, Ethics, and Humanities in Medicine, 7,* 4. https://doi.org/10.1186/1747-5341-7-4

8

Methods for Determining Deliberative Consensus

A number of specific methods have been used for determining deliberative consensus, each of which involves decisions about how expertise and consensus are defined. This chapter describes ten of these methods and gives examples of their use: Delphi studies, the nominal group technique, surveys of experts, systematic analysis of conclusions in peer-reviewed literature, consensus conferences, expert working groups, scientific citation networks, prediction markets, artificial intelligence and bespoke complex consensus methods. The following chapter (Chap. 9) will examine the adequacy of these methods for validly determining a scientific consensus.

8.1 Delphi Studies

The Delphi method was originally developed for use in forecasting (e.g. of future military threats to a country), but has also become a commonly used method for determining scientific consensus (Hsu & Sandford, 2019; Jorm, 2015). A Delphi study involves a number of steps. A group of individuals with expertise on some topic needs to be recruited. The

© The Author(s) 2025
A. Jorm, *Expert Consensus in Science*, https://doi.org/10.1007/978-981-97-9222-1_8

researchers compile a questionnaire with a list of statements that the experts anonymously and independently rate for agreement. The researchers then analyse the results and feed them back anonymously to the experts so that they can compare their ratings with the group. The experts are then able to revise their ratings after receiving the feedback. The responses converge across rounds of rating and are assessed against some statistical criterion for consensus. There are many variants of the Delphi method, which differ in how the initial statements for rating are produced, how the experts are recruited, whether the ratings are gathered in person or online, what type of feedback is given, how much discussion of reasons for rating is allowed and how many rounds of rating are used.

Delphi studies have been widely used in the health sciences to develop expert consensus on research methods, professional practice and scientific facts. Case Example 8.1 involves use of the Delphi method to develop consensus on research methodology, while Case Example 8.2 illustrates its use to arrive at a consensus on clinical practice. Case Example 7.5 (presented in Chap. 7) illustrates the use of the method to develop consensus on scientific facts, in this case the global prevalence of dementia.

Case Example 8.1 Use of Delphi Method to Produce Criteria for Quality Assessment of Randomized Controlled Trials

Systematic reviews are often used to summarize the findings from randomized controlled trials of health interventions. Such reviews typically include a methodological assessment of the quality of each of the trials in the review. However, to do this requires criteria for defining the quality of a trial. Verhagen et al. (1998) used the Delphi method to obtain consensus among experts about the core items that should be used for quality assessment. The researchers attempted to recruit a diverse group of experts, which included authors of publications on quality assessment, epidemiologists and statisticians. They created an initial questionnaire based on the items included in existing quality criteria lists for randomized controlled trials (e.g. Was a method of randomization performed? Was the outcome assessor blinded?). In the first round of the Delphi study, experts were asked to rate items on a five-point scale from "0: strongly disagree" to "4: strongly agree". Participants could suggest alternative wordings and new items about quality criteria. The researchers found substantial disagreement

(continued)

Case Example 8.1 (continued)

among the experts to the items in the first-round survey, with the statisticians differing from the authors and epidemiologists about what was important. The items were rated again in two subsequent rounds, which included modifications based on comments made by the experts. The researchers wanted a short final list of criteria, so they set a high threshold for consensus and only included items if they had a mean score of 2.8 or more on the 0–4 scale. From an initial pool of 206 items, the experts were able to reach consensus on nine quality criteria.

Case Example 8.2 Use of Delphi Method to Produce Guidelines on Sedation and Analgesia in Critically Ill Children

Because there was limited evidence from randomized controlled trials on analgesia and sedation in critically ill children, the UK Paediatric Intensive Care Society's Sedation, Analgesia and Neuromuscular Blockade Working Group carried out an expert consensus study to develop clinical practice guidelines using a variant of the Delphi method (Playfor et al., 2006). The Working Group was an interdisciplinary group of 13 intensive care unit personnel including physicians, nurses and pharmacists. An initial consensus conference produced a draft set of guidelines, which were sent to the 13 Working Group participants and independently rated for agreement on a nine-point scale ranging from "1: Disagree strongly" to "9: Agree strongly". Consensus was defined as 90% of the experts giving a rating of 7 or more on this scale. Recommendations that did not reach consensus on the first round of voting were rewritten to incorporate suggestions made. After three rounds of voting, recommendations that did not reach consensus were excluded. Of the 20 draft recommendations, all eventually achieved consensus and were included in the guidelines.

The Delphi method has also found widespread application in biological and social sciences. Case Example 8.3 illustrates a use in the biological sciences to develop a set of priorities for action, and Case Example 8.4 illustrates its use in the social sciences to develop advice for policymakers and the public.

Case Example 8.3 Use of Delphi Method to Determine Site Selection for European Oyster Habitat Restoration Projects

The European oyster is a threatened species, which was formerly common in European coastlines, but has been affected by overfishing and habitat destruction. Hughes et al. (2023) carried out a Delphi study to identify appropriate sites for habitat restoration projects. Twenty-five experts were recruited through the Native Oyster Restoration Alliance, which includes representatives from government agencies, science, non-government organizations, oyster growers and fisheries cooperatives. In the first-round questionnaire, the experts were asked to provide a list of potential selection factors. The researchers then categorized these factors and removed duplicates. In the second-round questionnaire, the experts were asked to rate each of 96 potential selection factors as "essential", "desirable", "not necessary" or "not sure". Consensus was defined as >70% of experts rating a factor as "essential" or "desirable". The experts were asked to review the findings from the second-round survey and given the opportunity to comment. Factors that did not reach the consensus criterion were reconsidered in a third and final round. Consensus was reached on 65 key factors, which covered abiotic, socio-economic and logistical factors.

Case Example 8.4 Use of Delphi Method to Identify Strategies for Improving Life Satisfaction

There is extensive research on improving life satisfaction, but this is complex to review. As an alternative, Buettner et al. (2020) used the Delphi method to assess the opinions of 20 senior researchers who were familiar with this evidence. The aim of the study was to find what policies are most likely to yield greater happiness for nations, and what individual strategies are most likely to enhance people's happiness. The investigators asked the experts to suggest ideas to address these two aims, which they sorted and reworded to eliminate redundancy. The experts were then asked to rate each of the strategies for effectiveness and feasibility on 1–5 scales, and the two ratings were summed to give scores from 2 to 10. The experts' ratings were examined for mean rating (reflecting level of endorsement) and standard deviation (reflecting degree of disagreement among experts). These results were fed back to the experts, and they were invited to make comments. In a second round of voting, the experts were asked to reconsider the strategies on which there was the most initial disagreement. The final list of strategies was made up of those with high average ratings and lower disagreement. The most strongly recommended policies involved investing in happiness research, strengthening social bonds, promoting good governance and investing in education. The recommended individual strategies were investing in social bonds, keeping learning and leading an active life.

8.2   Nominal Group Technique

A closely related method to the Delphi method is the Nominal Group Technique (Harb et al., 2021), which involves a structured group process for identifying problems, generating solutions and making decisions. The technique is carried out in a meeting with a facilitator who explains the question to be answered and the procedure to be followed. The facilitator asks participants to write down their ideas on the question. These ideas are then shared with the group without discussion. Next, the ideas are discussed by the group. Finally, members of the group independently rank or rate the ideas to determine the final outcome. As with Delphi, the Nominal Group Technique has many variants, with differences in how group members are selected, how the options for the group to consider are elicited, shared and refined, and how they are evaluated (Harb et al., 2021). This method has been used much less commonly than Delphi to determine scientific consensus. A major constraint is that it requires a face-to-face or online meeting, whereas Delphi studies are often done using online surveys which experts complete in their own time. Also, Nominal Groups can feasibly deal with a much smaller number of decisions than the Delphi method.

Most uses of the Nominal Group Technique for scientific consensus have been in the health sciences (Harb et al., 2021). Case Example 8.5 describes how it was used to determine a consensus on diagnosis of irritable bowel syndrome. Case Example 7.4 (see Chap. 7) also used the Nominal Group Technique to find messages suitable for use in campaigns to reduce stigma towards people with a mental illness.

Case Example 8.5 Use of the Nominal Group Technique to Determine Consensus on Diagnosis of Irritable Bowel Syndrome (IBS) in Primary Care

Patients with IBS present first to GPs, who are generally the sole providers of care. GPs therefore need to be able to reliably diagnose this condition. Diagnostic criteria for IBS have been developed for specialist medical settings, but these have not been as useful in primary care. Rubin et al. (2006) therefore used the Nominal Group Technique to develop a consensus on the diagnosis of IBS in primary care. The experts were ten GPs with a special

(continued)

Case Example 8.5 (continued)

interest in gastroenterology and two gastroenterologists, drawn from ten European countries. The researchers developed a series of 242 patient scenarios with different patterns of symptoms, signs, and risk factors, which were initially rated by the experts on a nine-point scale for level of agreement that the patient had IBS. The experts then met for two days to discuss the scenarios. They were provided with feedback on their own ratings and how they compared to the group's ratings, and they discussed each scenario in turn. This was followed by a confidential re-rating. Consensus that the person had IBS was defined as at least ten of the twelve experts scoring 7–9 on the nine-point rating scale. There was consensus about several defining symptoms of IBS, the duration of symptoms required for diagnosis, supporting risk factors and physical examinations that should be carried out.

8.3   Surveys of Experts

Formal surveys of members of scientific and professional societies or of authors of peer-reviewed publications on the topic of interest have sometimes been used to assess the consensus of experts. They are similar to the first round of a Delphi study, but no feedback is given to the experts on the results, and they have no subsequent rounds of rating. These methods have been extensively used in the area of human-caused climate change and show overwhelming support for its existence, as seen in Case Example 8.6.

Chapter 3 described the proposal of Peter Vickers in his book *Identifying Future Proof Science* (Vickers, 2023) that a statement could be regarded as a scientific fact if there was "a solid scientific consensus amounting to at least 95%, in a scientific community that is large, international, and diverse" (p. 18). To make this proposal concrete, Vickers has recently set up an Institute for Ascertaining Scientific Consensus, which is described in Case Example 8.7.

Previous chapters have presented other examples of the use of expert surveys to establish whether a consensus exists. Case Example 4.5 (see Chap. 4) described the use of a survey of published researchers to investigate the causes of the demise of the Neanderthals, which found support for the role of

Case Example 8.6 Use of Surveys to Assess Expert Consensus on Human-Caused Climate Change

Cook et al. (2016) reviewed surveys of scientists on human-caused climate change that have been carried out since 1991. The scientists were sampled from a variety of sources, including membership of professional bodies, signatories of public statements and conference attendees. The samples of scientists came from a range of countries. Questions asked also varied; examples include: "How convinced are you that most of recent or near future climate change is, or will be, a result of anthropogenic causes?" and "Do you think human activity is a significant contributing factor in changing mean global temperature?". Cook et al. divided the survey responses according to whether the sample of scientists included non-publishing climatologists or was restricted to publishing climatologists. They found a greater level of consensus among scientists with a higher level of expertise and concluded that "scientific consensus… is robust, with a range of 90%–100% depending on the exact question, timing and sampling methodology".

Case Example 8.7 Use of Online Surveys by the Institute for Ascertaining Scientific Consensus

Peter Vickers has set up an international network of scientists who can be polled on whether they agree or disagree with specific scientific statements (Adam, 2023). He has provisionally titled this network The Institute for Ascertaining Scientific Consensus (Institute for Ascertaining Scientific Consensus, 2024). At the time of writing, the Institute was still in its early stages, but the aim was to be able to email a global panel of over 100,000 scientists, who could respond to a specific question in under two minutes, thus placing minimal burden on participants. It was proposed that the first statement to be polled would be an uncontentious one which could test the methodology, such as "Science has put it beyond reasonable doubt that COVID-19 is caused by a virus". The Institute is a novel approach, but its feasibility remains to be seen.

demographic factors but disagreement about other possible causes. Similarly, Case Example 2.8 (Chap. 2) described how the International Astronomical Union used a vote of members at a conference in deciding to remove Pluto from the official list of planets.

8.4 Systematic Analysis of Conclusions in Peer-Reviewed Literature

The consensus of scientists can also be inferred by systematically analysing the conclusions they draw in peer-reviewed publications for level of consistency. The advantage of this approach over a survey is that a high level of expertise is ensured because all the experts have published on the topic. On the other hand, it may sometimes be difficult to infer a clear position on an issue from a publication, as it is common for researchers to draw tentative conclusions and to note the limitations of their work. This method has been used on the topic of climate change, as illustrated in Case Example 8.8.

Case Example 8.8 Consensus in the Scientific Literature on the Role of Human Activity in Global Warming

Cook et al. (2013) attempted to quantify the scientific consensus on the role of human activity in producing global warming by analysing the content of the peer-reviewed scientific literature. They searched the Web of Science database for papers on the topic published between 1991 and 2011 and found over 11,000. The researchers rated the abstracts of these papers for whether the authors made an endorsement of human activity as a cause of global warming, a rejection or took no position. Most abstracts took no position, but where a position was taken, 97% made an explicit or implied endorsement. The researchers also contacted authors of a sample of the papers and asked them to rate their own papers. Although the response rate was low (14%), the findings were similar, with 97% endorsement of human activity as a cause of global warming. More recently, Lynas et al. (Lynas et al., 2021) carried out a similar rating study on papers published since 2012. Again, they found that most papers took no position, but where the authors did take a position there was >99% endorsement. There have been criticisms of the methodology and conclusions of the Cook et al. (2013) study. Montford (2014), a climate change sceptic, has argued that the raters used by Cook et al. were climate activists and not independent. He also argued that what Cook et al. assessed was a "shallow consensus" that some proportion of global warming is due to human activity rather than a "deep consensus" that all or most of the warming is due to human activity. Montford claims that even climate sceptics would agree with the shallow consensus and that it is the deep consensus that is under debate.

8.5   Consensus Conferences

In the 1970s, the US National Institutes of Health (NIH) developed consensus conferences as a method of producing evidence-based statements on controversial medical issues. Over 160 consensus statements were produced before this approach was discontinued in 2013 (National Institutes of Health, 2023), covering such topics as lactose intolerance and health, preventing Alzheimer's disease, and hydroxyurea treatment for sickle cell disease. The consensus conference method was taken up by a number of other countries, usually in modified form (McGlynn et al., 1990). As implemented by NIH, a consensus conference involved assembling a panel of 9 to 16 members who were given the task of developing a consensus statement (Ferguson, 1996). The panel met in public session for presentation of data by invited experts. The panel prepared a draft consensus statement in private which was then presented in plenary session for discussion. The panel then redrafted, if necessary, and the statement was formally adopted as the output from the conference. Although the NIH has discontinued its consensus conferences, similar processes are still in use by other organizations worldwide.

Case Example 8.9 describes the process and outcomes of one of the NIH conferences. As in this example, consensus conferences do not report on details such as how panel members were selected, how consensus was arrived at and the level of consensus.

Case Example 8.9 Consensus Conference on Lactose Intolerance and Health

A consensus conference on "Lactose Intolerance and Health" was convened in 2010 by the Eunice Kennedy Shriver National Institute of Child Health and Development and the Office of Medical Applications of Research of the NIH (Suchy et al., 2010). The rationale for the conference was that many people in the United States avoid dairy products, which may lead them to have deficient intakes of calcium and vitamin D, and consequently increase risk for a range of diseases. A 14-member panel representing a range of

(continued)

Case Example 8.9 (continued)

clinical and public health disciplines was set up to convene the conference and produce a consensus statement. A systematic review of the literature was commissioned to inform the work of the panel. During the conference, the panel and conference audience heard evidence from 22 experts from relevant fields. Based on the systematic review and testimony of experts, the panel drafted a consensus statement and presented this for comment in an open forum on the final day of the conference. A final version of the consensus statement was released later that day. The conclusions in the statement included the fact that many people with lactose malabsorption do not have clinical lactose intolerance and, in most cases, do not need to completely avoid dairy consumption. A need was identified for educational programmes and behavioural approaches to improve the nutrition of these individuals.

8.6   Expert Working Groups

Expert working groups are often used by scientific and professional organizations to produce consensus statements. There is no standardized methodology for working groups, with variation in how issues for consensus are identified, how experts are chosen, how evidence is reviewed, how members work together and how they come to a consensus. There are many examples available of the use of working groups, including in the development of standards by the International Organization for Standardization (see Case Example 2.7, Chap. 2), in the standardization of nomenclature and measurement by the International Astronomical Union (see Case Example 2.8, Chap. 2), in the development of the International Classification of Diseases by the WHO (see Case Example 2.9, Chap. 2) and in the development of Sustainable Development Goals by the UN Development Programme (see Case Example 6.5, Chap. 6).

Case Example 8.10 illustrates the use of an expert working group to produce the State of Queensland's scientific consensus statement on the Great Barrier Reef, while Case Example 8.11 describes how the Australian and New Zealand College of Psychiatrists used a working group to produce clinical practice guidelines. The two examples provide a contrast in the degree of detail that the working groups provided about their processes for reaching consensus.

Case Example 8.10 Use of a Working Group to Develop the State of Queensland's Scientific Consensus Statement on the Great Barrier Reef

The Australian and Queensland governments have committed to a *Reef 2050 Water Quality Improvement Plan* to improve the quality of water flowing to the Great Barrier Reef. A 2017 Scientific Consensus Statement was produced to provide the scientific underpinning to the Plan (State of Queensland, 2017). The Scientific Consensus Statement was produced by a multidisciplinary group of scientists with expertise in Great Barrier Reef water quality science and management, who reviewed the available evidence. There were 11 "lead authors" and a larger group of "contributing authors". The report on the Consensus Statement provides no information on how the experts were chosen, how they went about reviewing the evidence, how they arrived at a consensus, what the level of consensus was, and whether there were dissenting voices. The Consensus Statement concluded that Great Barrier Reef ecosystems were in poor condition and that a major cause was the quality of water runoff associated with land development in the catchment area. The statement also concluded that current initiatives were insufficient to meet water quality targets and recommended further actions be taken.

Case Example 8.11 Use of a Working Group to Develop the Royal Australian and New Zealand College of Psychiatrists Clinical Practice Guidelines on Anxiety Disorders

In 2014, the Royal Australian and New Zealand College of Psychiatrists set a working group to develop clinical practice guidelines for Panic Disorder, Social Anxiety Disorder and Generalized Anxiety Disorder (Andrews et al., 2018). The working group consisted of eight healthcare academics and clinicians from Australia and New Zealand who were chosen to represent "a diverse range of expertise, opinion and adherence to particular therapeutic approaches". The working group had two face-to-face meetings and extensive email correspondence. For each diagnosis, a subgroup reviewed and summarized the evidence on treatment options. Clinical practice recommendations were developed by the whole group "through considerable frank and robust discussion to reach agreement". Levels of agreement were not reported, but the aim appeared to be 100% consensus. Drafts of the guidelines were circulated repeatedly among the eight members and areas

(continued)

Case Example 8.11 (continued)
of disagreement "resolved in an iterative process". A draft was also reviewed by eight national and international advisers and revised by the working group according to their suggestions. There was also a month-long period of public consultation where a draft was available on the College website and stakeholders were invited to make comments. The working group met by teleconference to consider all comments and recorded decisions on whether to revise the draft. There is no published account of what revisions occurred, if any. The final version of the guidelines was released in 2018.

8.7   Emerging Methods

The above methods are the ones most commonly used to establish consensus. However, there are several emerging methods that are seldom used at present but show promise for the future. These are scientific citation networks, prediction markets and artificial intelligence.

8.7.1 Scientific Citation Networks

Shwed and Bearman (2010) have proposed a method of assessing consensus on scientific facts using citation networks. The basis of this method is that citations to a publication most often indicate agreement with the conclusions of that publication. Where there is a lack of consensus, citation network communities form. These involve different groups of authors who preferentially cite each other. As consensus develops, the salience of these citation networks reduces: "When papers promote the same views and cite the same sources, the science behind them is conclusive" (p. 821). An advantage of this method is that it can be used to determine consensus without having to do an analysis of the claims made by each paper, which requires a high degree of content knowledge.

To validate this method, Shwed and Bearman (2010) applied it to cases that are now considered settled—that smoking causes cancer and that coffee does not. They then applied it to two cases in which there has been public controversy—the potential carcinogenicity of cell phones and the association between vaccines and autism. In both the cell phone and autism cases, they found a current scientific consensus supporting no association.

Bruggeman et al. (2012) agreed with Shwed and Bearman that most citations indicate agreement with the cited paper. However, they argue that it is necessary to also take account of the small number of cases where a paper is cited because the authors disagree with its conclusions. They showed that even if a small proportion of the citations indicate disagreement rather than agreement, it can alter the pattern of citation networks. Bruggeman et al. also disputed Shwed and Bearman's assumption that lack of citations across networks indicates that there is disagreement. They argue that lack of citations can simply indicate specialist sub-fields in a discipline, where specialists cite others from their subfield, without necessarily disagreeing with the scientists in other sub-fields.

Another limitation of this method is that it can be distorted by "citation cartels". These are groups of authors who collude to cite each other's work, so as to boost their own citation rankings or that of their institution. For example, such cartels have been found to operate in mathematics, where institutions in China, Saudi Arabia and Egypt started producing a large number of highly cited papers, whereas previously they had no record of excellence in the field (Catanzaro, 2024). Smaller fields of research such as mathematics are particularly vulnerable to such manipulation, because the extreme citing behaviour of a small number of individuals can have a disproportionate influence. Despite forming separate citation networks, these highly cited works do not indicate a lack of consensus in the field, but rather a deliberate manipulation of citing practices.

Although Shwed and Bearman's proposal to use citation network analysis to determine consensus has been around for well over a decade, it has not found widespread adoption. If Bruggeman et al. (2012) are correct that negative citations need to be distinguished from positive ones, this limits use of the method, as judgements have to be made about whether each citation indicates agreement or disagreement. The method is also potentially affected by deliberate attempts by a small number of individuals to manipulate citation counts.

8.7.2 Prediction Markets

Prediction markets allow participants to bet on the outcome of future events. They are applicable in situations where the betting is about an event which can be observed to have occurred or not at some future time. A common approach is to pay $1 if the event occurs and $0 if it does not. The predictions can be traded, and the prices indicate the collective forecast of the probability of the event.

It has been proposed that this method could be used in science to obtain consensus predictions of scientists (Hanson, 1995; Pfeiffer & Almenberg, 2010). A major limitation of using prediction markets in science is that they require a standard to determine who are the winners and losers of the bets. They could be used, for example, to predict events where there is an accepted standard measure, for example the coming year's economic growth, crop yield, mean ocean temperature or number of deaths from a disease. However, for complex scientific questions this is often not the case. More commonly, the eventual consensus of scientists with relevant expertise becomes the definer of truth, which creates a circular standard for predictions based on bets.

While this approach is limited in its scope for application in science, it has been used successfully to make predictions about the replicability of experiments in psychology and economics (Camerer et al., 2016; Dreber et al., 2015) and to predict national trends in infectious diseases (Li et al., 2016). Case Example 8.12 describes how a betting market was used to predict the replicability of psychology experiments.

Case Example 8.12 Use of a Prediction Market to Estimate the Reproducibility of Scientific Research

Psychological science has been described as having a "replicability crisis", where many published findings cannot be replicated by other researchers (Pashler & Harris, 2012). This has led some scientists to set up projects to try to systematically replicate influential findings. If there was some way of predicting which findings are less likely to replicate, this would allow replication efforts to be directed at those. To investigate whether this is possible, Dreber et al. (2015) set up a prediction market where scientists could

(continued)

Case Example 8.12 (continued)

bet on whether 44 studies published in prominent psychology journals could be replicated. The participants could trade contracts that paid $1 if a study was replicated and $0 if it was not. The 44 studies were repeated as part of the *Reproducibility Project: Psychology,* with a replication defined as P<.05 in the same direction as the original study. A market price of >$0.50 was regarded as predicting a successful replication, while one of <$0.50 predicted a failure to replicate. The researchers found that the prediction market correctly predicted 71% of the replications. Based on these results, Dreber et al. proposed that prediction markets could be used to prioritize efforts to replicate studies that have a low likelihood of replication.

8.7.3 Artificial Intelligence

Artificial intelligence could potentially automate the assessment of consensus in peer-reviewed literature. An app called *Consensus* (Consensus, 2024) provides an early attempt to do this. The app allows users to ask questions that are answered using data from over 200 million scientific papers. For example, when asked, "Do antidepressants work for anxiety disorders?" Consensus produced the following summary based on the top ten papers analysed:

> Most studies suggest that antidepressants are effective in treating anxiety disorders, while other studies argue that the benefits may be largely due to the placebo effect.

Some of the papers that the app drew on for this conclusion were old, including overviews for clinicians published in 1999 and 2000, and a meta-analysis published in 2003.

The app also provided a "Consensus Meter" which analysed 11 papers and concluded that 82% supported "Yes", 9% "Possibly" and 9% "No". The 11 papers used by the Consensus Meter represent only a small proportion of the available evidence. For comparison, a systematic review of randomized controlled trials examining the effects of two classes of antidepressants for treating anxiety found 135 relevant studies and concluded that all medications were more effective than placebo (Gosmann et al., 2021).

Although the use of artificial intelligence for assessing scientific consensus is in its infancy, it is certain that these methods will become more sophisticated in the future and they could eventually replace manual analysis of the literature.

8.8   Bespoke Complex Consensus Methods

A number of organizations have developed their own bespoke methods of determining scientific consensus. The most prominent of these is the Intergovernmental Panel on Climate Change (IPCC) (Harris, 2021). The IPCC organizes a large number of experts from around the world into working groups which produce draft reports on various aspects of climate change. Authors are required to record all scientifically valid perspectives and to note where any evidence is not consistent with the consensus view. These reports are reviewed by other scientists and then by government representatives. Because the IPCC operates under the auspices of the United Nations, governments review the drafts line by line and every part of a report must be agreed on. Further description of the IPCC consensus process is given in Case Example 4.4 (Chap. 4). The IPCC consensus processes have been the subject of criticism from some climate scientists. Curry and Webster (2013) have argued that the IPCC has downplayed the uncertainty of the data and that its consensus process is dominated by more confident scientists.

Other examples involving a bespoke complex consensus method are the Commissions of *The Lancet* journal, which provide recommendations from international panels of experts to change health policy or improve practice. The consensus processes of these Lancet Commissions are not standardized and not described in the published reports. The journal's Information for Authors (Lancet, 2022) simply states:

> Topics for The Lancet Commissions are selected by our editors, who work with academic partners to identify the most pressing issues in science, medicine, and global health with the aim of producing recommendations to change public policy or improve practice. Projects usually last 2–3 years, and author groups will represent a broad range of international expertise.

All Lancet Commissions are academic publications and are subject to the same rigorous peer review process as all other research papers published in our journals. (p. 6)

There have been over 100 Lancet Commissions to date, covering such topics as pollution and health, legal determinants of health, health effects of climate change, elimination of viral hepatitis and protecting the physical health of people with mental illness. Lancet Commission reports draw on the prestige of the *Lancet* journal, which has one of the highest impact factors in science. They are generally highly cited, which is itself an indicator of agreement from the academic community.

8.9 Evaluating Consensus Methods

This chapter has described the range of methods that have been used to determine scientific consensus and has considered some of their limitations. However, it has avoided discussion of whether these methods result in valid decisions. To do this requires an examination of the evidence on the conditions under which groups come to good-quality judgements, which is the subject of the next chapter.

References

Adam, D. (2023). The consensus projects. *Nature, 617*, 452–454.

Andrews, G., Bell, C., Boyce, P., Gale, C., Lampe, L., Marwat, O., et al. (2018). Royal Australian and New Zealand College of Psychiatrists clinical practice guidelines for the treatment of panic disorder, social anxiety disorder and generalised anxiety disorder. *Australian and New Zealand Journal of Psychiatry, 52*, 1109–1172. https://doi.org/10.1177/0004867418799453

Bruggeman, J., Traag, V. A., & Uitermark, J. (2012). Detecting communities through network data. *American Sociological Review, 77*, 1050–1063. https://doi.org/10.1177/0003122412463574

Buettner, D., Nelson, T., & Veenhoven, R. (2020). Ways to greater happiness: A Delphi study. *Journal of Happiness Studies, 21*(8), 2789–2806. https://doi.org/10.1007/s10902-019-00199-3

Camerer, C. F., Dreber, A., Forsell, E., Ho, T. H., Huber, J., Johannesson, M., et al. (2016). Evaluating replicability of laboratory experiments in economics. *Science, 351*, 1433–1436. https://doi.org/10.1126/science.aaf0

Catanzaro, M. (2024). Citation cartels help some mathematicians – And their universities – Climb the rankings. *Science, 383*(6682). https://doi.org/10.1126/science.zcl2s6d

Consensus. (2024). *Consensus: AI search engine for research*. Retrieved August 18, 2024, from https://consensus.app/#

Cook, J., Nuccitelli, D., Green, S. A., Richardson, M., Winkler, B., Painting, R., et al. (2013). Quantifying the consensus on anthropogenic global warming in the scientific literature. *Environmental Research Letters, 8*, 024024.

Cook, J., Oreskes, N., Doran, P. T., Anderegg, W. R. L., Verheggen, B., Maibach, E. W., et al. (2016). Consensus on consensus: A synthesis of consensus estimates on human-caused global warming. *Environmental Research Letters, 11*(4), 048002. https://doi.org/10.1088/1748-9326/11/4/048002

Curry, J. A., & Webster, P. J. (2013). Climate change: No consensus on consensus. *CAB Reviews, 8*(001). https://doi.org/10.1079/PAVSNNR20138001

Dreber, A., Pfeiffer, T., Almenberg, J., Isaksson, S., Wilson, B., Chen, Y., et al. (2015). Using prediction markets to estimate the reproducibility of scientific research. *PNAS, 112*, 15343–15347. https://doi.org/10.1073/pnas.1516179112

Ferguson, J. H. (1996). The NIH Consensus Development Program. The evolution of guidelines. *International Journal of Technology Assessment in Health Care, 12*(3), 460–474. https://www.ncbi.nlm.nih.gov/pubmed/8840666

Gosmann, N. P., Costa, M. A., Jaeger, M. B., Motta, L. S., Frozi, J., Spanemberg, L., et al. (2021). Selective serotonin reuptake inhibitors, and serotonin and norepinephrine reuptake inhibitors for anxiety, obsessive-compulsive, and stress disorders: A 3-level network meta-analysis. *PLoS Medicine, 18*(6), e1003664. https://doi.org/10.1371/journal.pmed.1003664

Hanson, R. (1995). Could gambling save science? Encouraging an honest consensus. *Social Epistemology, 9*(1), 3–33. https://doi.org/10.1080/02691729508578768

Harb, S. I., Tao, L., Peláez, S., Boruff, J., Rice, D. B., & Shrier, I. (2021). Methodological options of the nominal group technique for survey item elicitation in health research: A scoping review. *Journal of Clinical Epidemiology, 139*, 140–148. https://doi.org/10.1016/j.jclinepi.2021.08.008

Harris, R. (2021, June 29). Climate explained: How the IPCC reaches scientific consensus on climate change. *The Conversation*. https://theconversation.com/climate-explained-how-the-ipcc-reaches-scientific-consensus-on-climate-change-162600

Hsu, C., & Sandford, B. A. (2019). The Delphi technique: Making sense of consensus. *Practical Assessment, Research and Evaluation, 12*, 10. https://doi.org/10.7275/Pdz9-th90

Hughes, A., Bonacic, K., Cameron, T., Collins, K., da Costa, F., Debney, A., et al. (2023). Site selection for European native oyster (Ostrea edulis) habitat restoration projects: An expert-derived consensus. *Aquatic Conservation: Marine and Freshwater Ecosystems, 33*, 721–736. https://doi.org/10.1002/aqc.3917

Institute for Ascertaining Scientific Consensus. (2024). *Institute for Ascertaining Scientific Consensus*. Retrieved August 18, 2024, from https://iasc.awh.durham.ac.uk/

Jorm, A. F. (2015). Using the Delphi expert consensus method in mental health research. *Australian and New Zealand Journal of Psychiatry, 49*(10), 887–897. https://doi.org/10.2307/4473279210.1177/0004867415600891

Lancet. (2022). *Information for authors*. www.thelancet.com

Li, E. Y., Tung, C. Y., & Chang, S. H. (2016). The wisdom of crowds in action: Forecasting epidemic diseases with a web-based prediction market system. *International Journal of Medical Informatics, 92*, 35–43. https://doi.org/10.1016/j.ijmedinf.2016.04.014

Lynas, M., Houlton, B. Z., & Perry, S. (2021). Greater than 99% consensus on human caused climate change in the peer-reviewed scientific literature. *Environmental Research Letters, 16*(11), 114005. https://doi.org/10.2307/4473279210.1088/1748-9326/ac2966

McGlynn, E. A., Kosecoff, J., & Brook, R. H. (1990). Format and conduct of consensus development conferences. Multi-nation comparison. *International Journal of Technology Assessment in Health Care, 6*(3), 450–469. https://doi.org/10.1017/s0266462300001045

Montford, A. (2014). *Fraud, bias and public relations: The 97% 'consensus' and its critics*. https://www.thegwpf.org/content/uploads/2014/09/Warming-consensus-and-it-critics1.pdf

National Institutes of Health. (2023). *NIH consensus development program*. Retrieved February 21, 2023, from https://consensus.nih.gov/

Pashler, H., & Harris, C. R. (2012). Is the replicability crisis overblown? Three arguments examined. *Perspectives in Psychological Science, 7*(6), 531–536. https://doi.org/10.1177/1745691612463401

Pfeiffer, T., & Almenberg, J. (2010). Prediction markets and their potential role in biomedical research – A review. *Biosystems, 102*(2–3), 71–76. https://doi.org/10.1016/j.biosystems.2010.09.005

Playfor, S., Jenkins, I., Boyles, C., Choonara, I., Davies, G., Haywood, T., et al. (2006). Consensus guidelines on sedation and analgesia in critically ill children. *Intensive Care Medicine, 32*(8), 1125–1136. https://doi.org/10.1007/s00134-006-0190-x

Rubin, G., De Wit, N., Meineche-Schmidt, V., Seifert, B., Hall, N., & Hungin, P. (2006). The diagnosis of IBS in primary care: Consensus development using nominal group technique. *Family Practrice, 23*(6), 687–692. https://doi.org/10.1093/fampra/cml050

Shwed, U., & Bearman, P. S. (2010). The temporal structure of scientific consensus formation. *American Sociological Review, 75*(6), 817–840. https://doi.org/10.1177/0003122410388488

State of Queensland. (2017). *2017 consensus statement: Land use impacts on Great Barrier Reef water quality and ecosystem condition.* State of Queensland.

Suchy, F. J., Brannon, P. M., Carpenter, T. O., Fernandez, J. R., Gilsanz, V., Gould, J. B., et al. (2010). NIH consensus development conference statement: Lactose intolerance and health. *NIH Consensus and State-of-the-Science Statements, 27*(2), 1–27. https://www.ncbi.nlm.nih.gov/pubmed/20186234

Verhagen, A. P., de Vet, H. C., de Bie, R. A., Kessels, A. G., Boers, M., Bouter, L. M., & Knipschild, P. G. (1998). The Delphi list: A criteria list for quality assessment of randomized clinical trials for conducting systematic reviews developed by Delphi consensus. *Journal of Clinical Epidemiology, 51*(12), 1235–1241. https://doi.org/10.1016/s0895-4356(98)00131-0

Vickers, P. (2023). *Identifying future-proof science.* Oxford University Press.

9

How Wisdom-of-Crowds Research Can Help Improve Deliberative Consensus Methods

This book has presented numerous examples of where expert consensus is used in science. However, use of expert consensus processes does not guarantee that it will result in optimal decisions. Evidence is needed on the conditions under which expert consensus achieves this. Fortunately, there is an area of psychological science, which investigates group decision-making and can help validate consensus processes. There is extensive research showing that, under certain conditions, groups are more likely to make better-quality decisions than individual experts. In a best-selling book aimed at a general readership, James Surowiecki (2004) labelled this phenomenon "The Wisdom of Crowds", and this term has been adopted by many researchers in this field. Others refer to the field as "collective intelligence".

Wisdom-of-crowds research started with observations by Francis Galton (1907) about a competition held at an English livestock show. To quote Galton's account:

> A weight-judging competition was carried out at the annual show of the West of England Fat Stock and Poultry Exhibition recently held in Plymouth. A fat ox having been selected, competitors bought stamped and numbered cards, for 6d each, on which to inscribe their respective names,

© The Author(s) 2025
A. Jorm, *Expert Consensus in Science*, https://doi.org/10.1007/978-981-97-9222-1_9

addresses, and estimates of what the ox would weigh after it had been slaughtered and "dressed." Those who guessed most successfully received prizes. About 800 tickets were issued, which were kindly lent me for examination after they had fulfilled their immediate purpose. These afforded excellent material. The judgments were unbiassed by passion and uninfluenced by oratory and the like. The sixpenny fee deterred practical joking, and the hope of a prize and the joy of competition prompted each competitor to do his best. The competitors included butchers and farmers, some of whom were highly expert in judging the weight of cattle; others were probably guided by such information as they might pick up, and by their own fancies. (p. 450)

After removing 13 cards that were defective or illegible, Galton analysed the distribution of 787 guesses. These ranged from 1074 to 1293 lbs, with a median estimate of 1207 lbs, which was within 1% of the actual weight of 1198 lbs. In this case, the "crowd" was remarkably accurate.

More recent research has confirmed Galton's observation that aggregating information across a group of individuals (a "crowd") can result in impressive performance, even if the individuals are not selected for their expertise on the topic. Galton aggregated the group's judgements by using the median, but in other instance the group's mean or mode might be more appropriate, depending on how individual judgements are measured. Below are some illustrative studies examining crowd wisdom on a variety of tasks:

• *Fact-checking news headlines.* News items can sometimes be inaccurate. Fact-checkers have been employed to identify such misleading news items, but this is expensive to do. However, it has been found that the aggregated judgements of laypeople can be used to identify low-quality news sources and inaccurate news posts, which is a much cheaper alternative. Crowds of less than 20 people have been found to agree with fact-checkers as well as fact-checkers agree with each other (Martel et al., 2024). In one study, social media users from a range of countries were asked to rate the accuracy of COVID-19 news headlines (Arechar et al., 2023). Using fact-checking websites as the standard of truth, the researchers found that aggregated data from as few as 15 raters could differentiate true from false headlines over 90% of the time.

- *Forecasting election results.* Murr (2016) looked at citizen forecasts in seven British General Elections between 1964 and 2010. Voters in each constituency were asked: "Who do you think will win in your constituency?" Using a wisdom-of-crowds approach, Murr forecast that a constituency would be won by the party which most citizens said would win it. Across all elections, 61% of individuals made correct forecasts compared to 85% of crowds. Murr also looked at forecasted seat numbers in parliament compared with actual ones. There was a close match, with the mean absolute error for the three main parties being about eight seats. To give a concrete example, in the 2005 election, the wisdom-of-crowds' model forecasted 188 seats for the Conservatives, 379 for Labour and 63 for the Liberal Democrats. The actual numbers were 196, 366 and 64, respectively, giving a mean absolute error of about 8 seats. The wisdom-of-crowds model was also found to correctly predict all seven prime ministers.
- *Forecasting sporting match results.* O'Leary (2017) compared an online crowd with five soccer experts at predicting the outcomes of matches in the 2014 World Cup. The crowd were participants in Yahoo's "World Soccer Pick'em", in which individuals could attempt to pick the winner (or a draw) and the score. Yahoo assembled a group of five soccer experts to also make predictions, so it was possible to compare them to the crowd. The majority predictions of the Yahoo crowd were found to be correct in 45 out of 64 matches, which was better than the individual experts, who managed between 33 and 40 correct.
- *Answering quiz questions.* Simoiu et al. (2019) carried out an online experiment looking at crowd performance on 1000 questions across 50 topic domains, for example factual knowledge, popular culture, spatial reasoning and foreign language skills. They recruited 1707 online participants who gave individual answers. These were then aggregated to give crowd responses, using the median or the mode depending on whether the possible answers were quantitative (e.g. "In which year was this movie released?") or categorical (e.g. "What musical instrument is this?"). The researchers compared the crowd's scores on each of the topic domains to the distribution of individual scores

and found that the crowd beat 86% of individuals for quantitative questions and 87% for categorical questions.

In all these studies, the crowds are made up of individuals with no special expertise who are making independent judgements without conferring with other members of the crowd. The aggregation of their judgements is made by the researcher or an algorithm, without crowd involvement. Why do crowds do well on these tasks? In such situations, each individual has imperfect knowledge and is prone to error. However, by having a large number of individuals make judgements, these errors will tend to cancel out.

While crowds can often make wise decisions, they do not always do so. Wisdom-of-crowds research has suggested a number of conditions under which a crowd is more likely to make wise decisions (Larrick et al., 2012; Surowiecki, 2004). Four conditions which have consistent supporting evidence are discussed here: selection for expertise, cognitive diversity, independence of judgements and opportunity for sharing. These conditions have implications for how scientists should come to a deliberative consensus and are a standard against which methods for determining consensus can be validated.

9.1   Selection for Expertise

The wisdom-of-crowds phenomenon is based on the aggregation of judgements from individuals who may have limited expertise on the topic they are making judgements on. By contrast, scientific consensus processes involve individuals with a high level of expertise on a topic. It might be expected that crowds of experts would do better than crowds of amateurs, and this is indeed what the research shows. I illustrate this below with findings from two large studies.

Mannes et al. (2014) assembled 90 datasets where individuals made judgements that could be assessed for accuracy. There were 40 datasets from laboratory experiments where participants made numerical estimates (e.g. of temperatures, distances), with the number of participants varying from 15 to 413. The other 50 datasets were forecasts of economic indicators (e.g. the consumer price index and nominal gross domestic product) made by professional economists, with the median number of participants being 35. Mannes and colleagues compared the performance of the whole crowd with that of a select crowd made up of the five best performers on previous tasks, and also with the single best performer. In the experimental data, the select five-person crowd did best for 21 tasks, compared to 14 tasks for the whole crowd and 5 for the best member. For the economic data, the differences were even greater, with the five-person select crowd being most accurate for 34 forecasts, the whole crowd for 15 and the best member for only 1. These results show that quite a small crowd can do better than a large one if its members are selected for a high level of expertise. The findings also show that it is better to rely on a crowd than a top individual expert.

The second example concerns performance at forecasting real-world political events, such as "Will the Six-Party talks (among the US, North Korea, South Korea, Russia, China, and Japan) formally resume in 2011?" and "Who will be inaugurated as President of Russia in 2012? (a) Medvedev, (b) Putin, (c) Neither?" Mellers et al. (2014) recruited forecasters from a range of sources, with all required to have a bachelor's degree or higher. The forecasters were asked to provide predictions in two yearly rounds: 2011–2012 and 2012–2013. Mellers and colleagues looked at accuracy of forecasts for individuals and teams, both with and without training in how to make good forecasts. They also placed the highest 2% of performers from Year 1 (2011–2012) into elite "superforecaster" teams of 12 persons to see how good their forecasts were in Year 2 (2012–2013). While forming individuals into teams and training them improved the accuracy of forecasts, the largest effect by far was forming elite teams. Performance was measured by Brier scores, which can range from 0 (the best score) to 2 (the worst). In Year 2, average scores were 0.25 for individual forecasters, 0.16 for team forecasters and 0.07 for the elite team forecasters. The authors concluded that "the pooling of top performers into elite teams with the exalted title of 'superforecasters' was the equivalent of a 'steroid injection'… and far exceeded our wildest expectations" (p. 1113). Like the Mannes et al. (2014) study, this example shows that even quite small crowds of individuals with very high expertise can outperform much larger crowds of variable expertise.

While deliberative consensus processes involving scientists necessarily involve selection for relevant scientific expertise, the level of expertise can vary. Consensus among those with greater expertise can be given higher credibility. In the area of climate change, for example, scientists with a greater number of relevant publications and citations were found to have a higher consensus about anthropogenic effects on the climate (see Chap. 7, Case Example 7.3; and Chap. 8, Case Example 8.6).

9.2   Cognitive Diversity

"Cognitive diversity" refers to "the various ways in which people use and store information, including knowledge, beliefs, ability, expertise, goals, or values" (Sulik et al., 2022, p. 752). This has been distinguished from "surface" or "sociodemographic diversity", which involves factors such as gender, age and ethnicity.

Statistical and computational modelling of decision-making by crowds shows that better decisions result when the crowd is cognitively diverse, with its members' judgements being as negatively correlated with each other as possible, while still having a high level of expertise (Davis-Stober et al., 2014; Page, 2007). This result is not intuitively obvious and has been explained by Davis-Stober et al. (2014) using an analogy:

> A helpful analogy is to think of a group like a financial portfolio whose members are assets. It is useful to hedge one's bets by holding some assets that are negatively correlated with the rest of the portfolio, so that there are some positive returns when other assets perform poorly. Similarly, we find that wise groups should include some judges who predict better when others falter. (p. 97)

A situation where this could occur is where a decision task requires multiple types of knowledge to make a good judgement and no member of the crowd has all this knowledge. A crowd in which members complement each other in the type of knowledge they provide would do better than one in which the members share the same knowledge.

Another way to think about the role of diversity is to consider the opposite pattern of crowd composition, where the members' judgements are positively correlated with each other. In this situation, the errors of the individuals in the crowd may not cancel each other out. In the extreme case, where the individuals are perfectly correlated, they are clones of each other, and their judgements will be no better than that of a single individual.

Empirical studies also support a role for cognitive diversity. Keck and Tang (2020) looked at cognitive process diversity experimentally using three different tasks: providing estimates of dates of different historical events, probabilistic forecasting of the outcomes of soccer games and estimating the heights of individuals from a photograph. To vary cognitive strategies, they either asked participants to make their estimates using an analytic approach, using an intuitive approach, or gave them no specific instructions. The participants made independent judgements and were formed by the researchers into crowds. Crowds made up of participants with a diversity of strategies (analytic and intuitive) did better than groups with only one strategy or groups from the control condition.

Shi et al. (2019) looked at a different type of cognitive diversity—conservative versus liberal political values. They examined the quality of Wikipedia pages which were produced by an ideological diverse group of editors compared to a homogeneous group. The political ideology of an editor was judged by their contribution to conservative versus liberal Wikipedia articles, while the quality of an article was determined by Wikipedia's six-category quality scale, which ranges from Featured Article (highest) to Stub (lowest). The quality of articles was assessed in political, social issues and science areas. After adjusting for length of article, number of editors and number of edits, involvement of ideologically diverse editors increased the odds of a higher quality article by over 18 times for political articles and around 2 times for social issues and science articles. Shi et al. (2019) looked at why ideological diversity produced better-quality articles and concluded that "frequent, intense disagreement within ideologically polarized teams foments focused debate and, as a consequence, higher-quality edits that are more robust and comprehensive" (p. 334).

Can sociodemographic diversity in crowds play a similar role? To investigate this, de Oliveira and Nisbett (2018) looked at the effects of crowd composition on factors like age, gender and ethnicity on a variety of judgement tasks. They found very small effects, even when crowds that were homogeneous on several sociodemographic factors were compared to very diverse crowds. They concluded that sociodemographic diversity will only make a difference when the sociodemographic factor is at least moderately correlated with cognitive diversity.

While cognitive diversity can make a difference to quality of crowd decisions, the effect is seen more on some tasks than others. Sulik et al. (2022) concluded that the effect of diversity is most evident on complex problem-solving tasks and that these are often found in science. They state:

> Science is a prime example of a complex problem... It comprises multiple and various subtasks; it involves a rugged epistemic landscape; and it frequently includes conflicting perspectives, creativity, and problem posing. It also highlights how the promise of diversity can be diminished by institutional or individual bias. The role of diversity in science reflects in microcosm the role of diversity in general. (p. 761)

If science is to benefit from cognitive diversity, scientists need to be alert to issues for which diversity may be restricted. Staddon (2018) has pointed out that the growth of science has led to increasing specialization, with each sub-specialty producing its own journals. He has argued that some sub-specialties have developed in the social sciences which lack broader credibility. Submissions to these journals are reviewed and articles read only by like-minded persons. He gives as an example the area of Whiteness Studies, which rejects objectivity and the empirical standards of broader disciplines. The same point applies to pseudoscience areas like homeopathy, which also have their own specialist journals and are isolated from the broader medical sciences.

Feminist philosophers of science have argued that there can be "epistemic benefits" of including under-represented groups in science for areas of research where their different life experiences, values and interests are relevant. For example, Intemann (2009) has argued that the kind of diversity that is important to achieving epistemic benefits in particular research contexts may depend on the nature of the research. Projects that deal with human subjects may require a kind of researcher diversity not required by other research projects. If subject responses are influenced by the race and gender of researchers, it will be important to have a pool of researchers whose diversity corresponds to the kind of diversity in the subject pool, to the extent possible. In other research contexts, diversity of life experiences will be more epistemically salient. Research on issues that have global implications such as climate change, nanotechnology, or genetically modified food might benefit from researchers with geographical diversity (such as those who have lived in developing countries, as well as developed countries). Research on race and sex differences might be more objective with researchers from diverse social positions relevant to identifying the presence of stereotypes. Research on water quality on Sioux reservations could achieve epistemic benefits from the participation of Sioux researchers, or those with experiences living on a reservation. Participation of researchers with diverse political values might be important for assessing the risks related to levels of environmental toxins. Thus, the sort of diversity that is important to increasing objectivity in a particular case can depend on the content of the research. (p. 262)

However, for other areas of research, such as theoretical physics, where the diversity of human experiences is not relevant to the subject matter, having diverse researchers would not have any epistemic benefits (Intemann, 2009).

Concern has also been raised about the need for viewpoint diversity in psychological science (Frisby et al., 2023). Duarte et al. (2015) have pointed out that academic psychology lacks political diversity, with most members having left-leaning attitudes, and that this bias has increased in recent decades. They argue that this lack of diversity "can undermine the validity of social psychological science via mechanisms such as the embedding of liberal values into research questions and methods, steering researchers away from important but politically unpalatable research topics, and producing conclusions that mischaracterize liberals and conservatives alike" (p. 1). This lack of political diversity also affects other social science disciplines (Haidt & Jussim, 2016).

While many scientific consensus processes have endeavoured to involve a large and diverse group of scientists, others have been criticized for the lack of cognitive diversity in the expert group. Case Example 9.1 describes a consensus process where there were subsequent concerns about lack of diversity.

Case Example 9.1 Concerns About Lack of Diversity Among the Developers of the American Psychological Association's Practice Guidelines for Men and Boys

The American Psychological Association (2018) has produced Practice Guidelines for Men and Boys. The writing group that produced the guidelines was drawn from members of a division of the APA devoted to Psychological Study of Men and Masculinities. The guidelines give major attention to socialization practices with males and how these contribute to a variety of psychological and social problems in men and boys, and to lack of help-seeking for these problems. Following the publication of the guidelines, a number of critical responses appeared complaining that they were based on a strongly sociocultural perspective on masculinity and ignored biological influences. Reviewing these critiques, Ferguson (2023) concluded:

the guidelines failed to acknowledge significant evidence for biological influences on gender (e.g., hormonal, and hypothalamic influences on gender identity and gendered behavior), were unintentionally disparaging of traditional men and families, and were too closely wedded to specific sociocultural narratives and incurious of data not supporting those narratives. (p. 1)

He criticized the make-up of the writing group, stating that they "failed to reach out to conservative stakeholders and men whose worldviews could reasonably be expected to differ from those of liberal/progressive psychologists" (Ferguson, 2023, p. 6), and advocated for a revision of the guidelines to incorporate more diverse perspectives.

Another example concerns the task force that developed the third edition of the *American Psychiatric Association's Diagnostic and Statistical Manual of Mental Disorders* (*DSM-III*) (American Psychiatric Association, 1980), which has been criticized for being a "small, culturally-homogeneous subset of mental health professionals who were socially positioned at a given moment in psychiatric history to have their judgements ratified by the institutional apparatus of the APA" (Davies, 2017, p. 44). Such examples indicate the importance of considering cognitive diversity when planning scientific consensus activities.

The cognitive diversity of a group needs to be considered in combination with selection for expertise. There is a delicate line to be walked here, as cognitive diversity must not compromise a high level of expertise. Having a group of scientists that were diverse in level of expertise rather than type of expertise would not be likely to result in better judgements.

9.3   Independence of Judgements

The wisdom-of-crowds effect first observed by Galton (1907), and confirmed by many others, is based on the aggregation of data from individuals who have made independent judgements. Independence in this case means that the judgements of individuals are not determined by the judgements of those around them. If the judgements are not independent, then errors may not cancel each other out. Instead, judgements may be systematically biased in one direction. Where the members of a group are able to influence each other's judgements, biases may arise through "groupthink" and "herding".

Groupthink occurs when the individual members of a small cohesive group accept a conclusion that they think represents the group consensus, even if they or other members of the group do not personally accept the consensus. The concept of groupthink was proposed by social psychologist Irving Janis (1972) based on an analysis of how poor foreign policy decisions came to be made (e.g. the US decision to support the Bay of Pigs invasion of Cuba in 1961). Janis concluded that groupthink occurred only under certain unusual conditions, but subsequent research has indicated that it is a much more general effect in group decision-making (Baron, 2005). Baron (2005) described the common experience of groupthink as follows:

> most of us, I suspect, have been in settings in which our private reservations regarding some group option have been assuaged by a seeming con-

sensus of our group mates or where our concerns about having pleasant social interactions and our own social acceptance take precedence over any need to explore every last objection and nuance to a collective decision. (p. 227)

Baron (2005) proposed that groupthink is most likely to occur when the members have a strong identification with the group, where group interaction and discussion begin to produce a group norm, and where the members lack confidence in their ability to come to a satisfactory decision.

Herding is a similar concept and involves the tendency to copy other people's behaviours. It is seen in situations where individuals in the crowd make their decisions known to others over a period of time, so that early decisions can influence later ones. In financial markets, it can lead to extreme phenomena like bubbles and crashes. Herding can be seen in online crowd behaviour. This was investigated in an experiment by Muchnik et al. (2013) using a social news aggregation website. Users of this site could post articles and then other users could comment on these. Comments could be "up-voted" or "down-voted" by other users. When posting a vote on a comment, users could see the current score for that comment and potentially be influenced by it. In this experiment, the researchers randomly gave either an initial up-vote or a down-vote and observed the subsequent behaviour of other users. They found that initial positive votes tended to influence other users and lead to a positive ratings bubble, increasing the final ratings by an average of 25%, whereas negative votes tended to get neutralized by crowd correction.

A number of other studies illustrate how independence can be compromised by knowing the responses of others in the group, leading to poorer group decisions. Earlier in this chapter, I described an online experiment by Simoiu et al. (2019) looking at crowd performance on 1000 questions across 50 topic domains. These researchers found a wisdom-of-crowds effect, with the aggregated performance of the crowd beating 86–87% of individuals. However, they also looked at what happened if the individuals were given information about the judgements of the group (most frequent answers or median response up to that point) before they made their own decisions. When this was done, the performance of the crowd decreased from the 86–87th percentile to the 80–81st percentile.

Frey and Van de Rijt (2021) had similar findings in a study where individuals worked independently at computer stations to answer quiz questions like "What was the year of the German invasion of Denmark?" When the participants answered sequentially and were informed about how prior participants had responded, the group tended to do worse than if they answered independently, particularly with harder questions.

Toyokawa et al. (2019) also found that herding was greater in more challenging tasks. They did an interactive online experiment with a gambling task called the "three-armed bandit", in which participants had to choose which of three slot machines to play, with the machines varying in their probability of payoff. The task could be made more difficult and the optimal response more uncertain by abruptly changing the payoffs of the three slot machines. The influence of the crowd was investigated by giving individuals feedback on how frequently other people were choosing the three slots. As the difficulty of the gambling task increased, individuals became more likely to conform to what others were deciding, particularly if there was a large crowd.

Another factor undermining independence is the influence of a powerful individual. Locke and Anderson (2015) did a series of experiments where they had pairs of individuals work together on decision-making tasks, such as selecting the best candidates for a job. One member of each pair was a confederate of the researchers who was trained in how to interact with the participant. When a participant worked with a confident powerful person, they participated less and deferred more, even when that person was wrong.

To summarize these studies, the independence of judgements which underlies the wisdom-of-crowds effect can be undermined where the decisions of individuals are announced to the crowd sequentially, particularly where the decision is a difficult one involving considerable uncertainty, and where a powerful individual has the opportunity to influence the crowd.

Independence of judgements is clear in some scientific consensus processes (e.g. Delphi studies and surveys of experts) where individual participants make private decisions which are not fed back to the group until

Case Example 9.2 Threats to Independence in the Royal Australian and New Zealand College of Psychiatrists Clinical Practice Guidelines on Schizophrenia

These clinical practice guidelines were developed by an expert working group consisting of eight psychiatrists, one clinical psychologist and one pharmacist (Galletly et al., 2016). Individual members drafted sections of the guidelines in their areas of interest and expertise. The group discussed the drafts in teleconferences, and, if there was disagreement, the issue was discussed until consensus was reached. Whether or not the working group members made independent judgements is in hindsight unknowable. However, this method could potentially be subject to the influence of strong individuals who wrote the drafts, dominance of the working group by one profession (psychiatrists) and group pressure for unanimity. The development of these guidelines would have involved greater independence if there was anonymity of the authors of the drafts, of comments by other group members and of voting on the final version.

Examples of more serious violations of both the independence and diversity principles are found in formal statements against a role for human activity in climate change, which have been produced by scientists who were recruited specifically because of their contrarian positions (see Chap. 7, Case Example 7.2).

9.4   Opportunity for Sharing

While independence is important to the wisdom-of-crowds effect, there is also evidence that sharing of information and discussion among members of the crowd can improve judgements in some circumstances (Becker et al., 2017; Dezecache et al., 2022; Granovskiy et al., 2015; Gürçay et al., 2015; Mellers et al., 2014; Mercier & Claidière, 2022; Navajas et al., 2018; Rowe et al., 2005). For these benefits to be seen, the sharing needs to occur in ways that minimize biases due to groupthink and herding. In the studies that show benefits, individuals are asked to provide independent judgements, then there is a sharing of information or discussion, followed by another round of independent judgements which are aggregated to give the final crowd judgement. This type of sharing is very different from what commonly occurs when groups make decisions. In a typical group decision, the individuals in the group are not asked to make an initial independent judgement before sharing, and individual judgements are declared in an identifying way rather than anonymously. In such circumstances, powerful and confident individuals may have a greater influence, there is pressure to conform, members may be reluctant to share their knowledge and dissenting opinions may be withheld (Sunstein, 2006). Below I describe some typical studies showing the conditions under which sharing of information and discussion produce benefits to crowd decisions.

One type of sharing is to simply provide feedback to individuals about the judgements of the crowd before seeking another round of independent judgements. This type of sharing was investigated in an experiment by Becker et al. (2017) in which participants were recruited via the World Wide Web to be players in an "Intelligence Game". Participants had to make estimations (e.g. they were shown a picture of food and asked to estimate the number of calories, or a picture of a jar of candies and asked to estimate the number of candies). The participants were randomly assigned to one of two conditions where they received feedback on other participants' judgements or to a control condition where they received no feedback. In the feedback conditions, they were placed in a "decentralized network" where all members of the crowd were equally connected with each other for feedback, or a "centralized network" where one member had a high number of connections to others, giving them a disproportionate influence on the feedback provided. After feedback, the members of the decentralized networks improved in the accuracy of their judgements. However, with the centralized networks, the central individual dominated the estimation process and group estimates were worse if this individual made an inaccurate initial estimate. These findings show that feedback from the crowd can produce better judgements where the conditions reduce dominance by a particular individual. Granovskiy et al. (2015) similarly found that feedback on the judgements of the crowd improved group performance on estimation tasks. This occurred through error correction as individuals who were outliers with answers far from the correct answer changed their judgements to be closer to the group mean.

Other studies have looked at the effects of discussion on accuracy of crowd judgements. A typical experiment was carried out by Navajas et al. (2018). They did research with a crowd of over 5000 people attending a popular live event. A speaker on the stage asked attendees eight questions involving general knowledge estimates (e.g. "What is the height in metres of the Eiffel Tower?"). Participants were provided with a pen and answer sheet linked to their seat number. After providing their independent answers, the participants were told to organize into groups of five based on a code on their answer sheet. The speaker repeated four of the eight questions and asked the groups to provide a consensus answer. Finally, the eight questions were asked again by the speaker and participants gave their revised independent answers. The researchers first compared the aggregated estimates from the whole crowd with the aggregated consensus answers of the small groups—what the researchers called a "crowd of crowds". They found that the average of only four consensus answers from the small groups was more accurate than the average of the whole crowd. They also found that the revised answers of the whole crowd after discussion were more accurate than the initial estimates. The importance of these findings is that it is not necessary for a group of over 5000 individuals to have a group discussion involving everybody. Indeed, such a discussion would not be feasible. Instead, the crowd can be divided into small groups which have the discussion and then the consensus results of the small groups aggregated over the large crowd. Other studies on the effects of having discussions have found similar benefits to crowd judgements (Dezecache et al., 2022; Gürçay et al., 2015; Mellers et al., 2014; Mercier & Claidière, 2022).

9.5   Validity of Methods for Deliberative Consensus

The wisdom-of-crowds literature can be used to assess whether a deliberative consensus process is more or less likely to come to a good decision. Summarizing the above evidence and generalizing it to scientific consensus, a better process is one where:

1. The scientists selected for the consensus are an elite group in their field, rather than a broader sample of scientists in the discipline.
2. The scientists are diverse in their areas of specialist knowledge, disciplinary training, methodological expertise and values.
3. The methods for arriving at a consensus are designed to reduce the influence of strong voices and pressures for conformity, for example with independent and anonymous voting.
4. The experts have opportunities to share their expertise and judgements through anonymous feedback on group judgements and open discussion of reasons for judgements.

Research teams should consider each of these criteria when designing their own consensus process. However, in many cases, researchers may want to consider using one of the commonly used consensus methods reviewed in Chap. 8. How well these meet the criteria will depend on the details of implementation, but some general assessment of validity is possible, as summarized in Table 9.1.

None of the methods meet all of these four criteria. None of them require selection of an elite group of experts (Criterion 1), nor do they require selection for cognitive diversity (Criterion 2). How well these criteria are met depends on the details of the expert selection for a particular consensus process. Where the methods can be more clearly distinguished is in independence of judgements (Criterion 3) and opportunity for sharing (Criterion 4). Delphi studies and the nominal group technique meet both Criteria 3 and 4, whereas surveys of experts lack opportunity for sharing, and consensus conferences and expert working groups do not involve independence of judgements.

Table 9.1 Assessment of whether deliberative consensus methods meet the criteria for wise crowd judgements

Consensus method	Selection for Expertise	Cognitive diversity	Independence of judgements	Opportunity for sharing
Delphi studies	Depends	Depends	Yes	Yes
Nominal Group Technique	Depends	Depends	Yes	Yes
Surveys of Experts	Depends	Depends	Yes	No
Consensus conferences	Depends	Depends	No	Yes
Expert working groups	Depends	Depends	No	Yes
Systematic analysis of conclusions in peer-reviewed literature	Yes	Yes, if there is no publication bias	Yes, if there is no publication bias	Yes, if there is no publication bias

Systematic analysis of conclusions in the peer-reviewed literature is somewhat different from the other methods for deliberative consensus, because the researchers who are assessing consensus do not have control over potential biases that may occur when scientists carry out and publish research. They have to assume that the studies that have been published come from a cognitively diverse set of scientists, that the scientists are coming to independent conclusions and that there is open sharing and discussion of findings in the field. These criteria are less likely to be met if there is selective publication of positive findings or suppression of findings that do not support dominant ideologies, which can sometimes occur (Clark et al., 2023; Frisby et al., 2023).

9.6 Comparison with Views of History, Philosophy and Sociology of Science (HPSS) Scholars

Chapter 3 examined the range of positions that HPSS scholars have taken on whether scientific consensus can indicate truth. Despite apparent differences, there were areas of agreement about the conditions under which consensus is more likely to indicate scientific truth. I identified four conditions that were supported by multiple HPSS scholars:

1. The consensus needs to be rational, empirical and critically examined.
2. The group coming to the consensus needs to be diverse.
3. The group needs to be open-minded and there is no coercion of dissenters.
4. The group needs to be sufficiently large to get reliable results.

In proposing such conditions, HPSS scholars have made their cases by argument with limited recourse to data, apart from analysis of historical examples. When one scholar's arguments are put up against another's, it is possible that one might concede the superiority of the other's arguments and change their position. However, this is not evident in the HPSS literature, leading to an impasse in areas of disagreement. By contrast, the wisdom-of-crowds literature gives an empirical basis for judging the conditions that are likely to result in good group decisions, although admittedly this literature is based on judgement tasks that are generally much simpler than those facing scientists. Nevertheless, it is interesting to compare the HPSS scholars' conditions from Chap. 3 with the conclusions from wisdom-of-crowds research reviewed in this chapter.

Table 9.2 gives a comparison of the two sets of conditions. It can be seen that three of the HPSS scholars' conditions are broadly supported by wisdom-of-crowds research. These are:

• The consensus needs to be rational, empirical and critically examined.
• The group coming to the consensus needs to be diverse.
• The group needs to be open-minded and there is no coercion of dissenters.

Table 9.2 Comparison of HPSS scholars' conditions for consensus to be an indicator of scientific truth with the findings from wisdom-of-crowds research

Conditions proposed by HPSS scholars	Relevant findings from wisdom-of-crowds research
The consensus needs to be rational, empirical and critically examined	The experts have opportunities to share their expertise and judgements through anonymous feedback on group judgements and open discussion of reasons for judgements.
The group coming to the consensus needs to be diverse	The scientists are diverse in their areas of specialist knowledge, methodological expertise and values.
The group needs to be open-minded, and there is no coercion of dissenters	The methods for arriving at a consensus are designed to reduce the influence of strong voices and pressures for conformity, for example with independent and anonymous voting.
The group needs to be sufficiently large to get reliable results	
	The scientists selected for the consensus are an elite group in their field, rather than a broader sample of scientists in the discipline.

However, one condition proposed by some HPSS scholars (viz. the group needs to be sufficiently large to get reliable results) is not a conclusion from the wisdom-of-crowds literature. Certainly, from a statistical point of view, a larger sample will give a more precise estimate of a population value. If, for example, we wanted to find out whether at least 95% of a relevant group of scientists agree with a proposition, it would be desirable to have a sufficiently large sample to reliably distinguish a value of 95% from a value of 90%. However, the wisdom-of-crowds literature shows that comparatively small crowds can produce wise judgements if they meet the conditions of selection of an elite group, cognitive diversity, independence and opportunity for sharing.

There is also one condition in the wisdom-of-crowds literature (viz. selection of an elite group), which does not appear in HPSS literature. However, it may be that this is an implied premise in the concept of expert consensus and does not require philosophical justification.

In drawing these conclusions from wisdom-of-crowds research, there is necessarily a leap of generalization from the often-trivial decision tasks used in this research and the much more complex consensus decisions that scientists make. There is a major difference in complexity from decisions about "How many candies are in a jar?" or "How high is the Eiffel Tower?" to estimating the contribution of human activity to climate change. Presently, wisdom-of-crowds research has not studied decisions on realistic scientific judgement tasks. However, it may be possible to develop such an area of research, which is the subject of the next chapter—the wisdom of scientific crowds.

References

American Psychiatric Association. (1980). *Diagnostic and statistical manual of mental disorders* (3rd ed.). Washington, DC: Author.

American Psychological Association. (2018). *APA guidelines for psychological practice with boys and men*. American Psychological Association. https://www.apa.org/about/policy/boys-men-practice-guidelines.pdf

Arechar, A. A., Allen, J., Berinsky, A. J., Cole, R., Epstein, Z., Garimella, K., et al. (2023). Understanding and combatting misinformation across 16 countries on six continents. *Nature Human Behaviour, 7*(9), 1502–1513. https://doi.org/10.1038/s41562-023-01641-6

Baron, R. S. (2005). So right it's wrong: Groupthink and the ubiquitous nature of polarized group decision making. *Advances in Experimental Social Psychology, 37*, 219–253. https://doi.org/10.1016/S0065-2601(05)37004-3

Becker, J., Brackbill, D., & Centola, D. (2017). Network dynamics of social influence in the wisdom of crowds. *Proceedings of the National Academy of Sciences USA, 114*(26), E5070–E5076. https://doi.org/10.1073/pnas.1615978114

Clark, C. J., Jussim, L., Frey, K., Stevens, S. T., Al-Gharbi, M., Aquino, K., et al. (2023). Prosocial motives underlie scientific censorship by scientists: A perspective and research agenda. *Proceedings of the National Academy of Sciences U S A, 120*(48), e2301642120. https://doi.org/10.1073/pnas.2301642120

Davies, J. (2017). How voting and consensus created the Diagnostic and Statistical Manual of Mental Disorders (DSM-III). *Anthropology & Medicine, 24*, 32–46. https://doi.org/10.1080/13648470.2016.1226684

Davis-Stober, C. P., Budescu, D. V., Dana, J., & Broomell, S. B. (2014). When is a crowd wise? *Decision, 1*, 79–101. https://doi.org/10.1037/dec0000004

de Oliveira, S., & Nisbett, R. E. (2018). Demographically diverse crowds are typically not much wiser than homogeneous crowds. *Proceedings of the National Academy of Sciences U S A, 115*(9), 2066–2071. https://doi.org/10.1073/pnas.1717632115

Dezecache, G., Dockendorff, M., Ferreiro, D. N., Deroy, O., & Bahrami, B. (2022). Democratic forecast: Small groups predict the future better than individuals and crowds. *Journal of Experimental Psychology: Applied, 28*(3), 525–537. https://doi.org/10.1037/xap0000424

Duarte, J. L., Crawford, J. T., Stern, C., Haidt, J., Jussim, L., & Tetlock, P. E. (2015). Political diversity will improve social psychological science. *Behavioral and Brain Sciences, 38*, e130. https://doi.org/10.1017/S0140525X14000430

Ferguson, C. J. (2023). The American Psychological Association's practice guidelines for men and boys: Are they hurting rather than helping male mental wellness? *New Ideas in Psychology, 68*, 100984. https://doi.org/10.1016/j.newideapsych.2022.100984

Frey, V., & Van de Rijt, A. (2021). Social influence undermines the wisdom of crowds in sequential decision making. *Management Science, 67*, 4273–4286. https://doi.org/10.1287/mnsc.2020.3713

Frisby, C. L., Redding, R. E., O'Donoghue, W. T., & Lilienfeld, S. O. (Eds.). (2023). *Ideological and political bias in psychology*. Springer.

Galletly, C., Castle, D., Dark, F., Humberstone, V., Jablensky, A., Killackey, E., et al. (2016). Royal Australian and New Zealand College of Psychiatrists clinical practice guidelines for the management of schizophrenia and related disorders. *The Australian and New Zealand Journal of Psychiatry, 50*(5), 410–472. https://doi.org/10.1177/0004867416641195

Galton, F. (1907). Vox populi. *Nature, 75*(1949), 450–451. https://doi.org/10.1038/075450a0

Granovskiy, B., Gold, J. M., Sumpter, D. J., & Goldstone, R. L. (2015). Integration of social information by human groups. *Topics in Cognitive Science, 7*(3), 469–493. https://doi.org/10.1111/tops.12150

Gürçay, B., Mellers, B. A., & Baron, J. (2015). The power of social influence on estimation accuracy. *Journal of Behavioral Decision Making, 28*, 250–261. https://doi.org/10.1002/bdm.1843

Haidt, J., & Jussim, L. (2016). Psychological science and viewpoint diversity. *Observer, 28*(2).

Intemann, K. (2009). Why diversity matters: Understanding and applying the diversity component of the National Science Foundation's broader impacts criterion. *Social Epistemology, 23*, 249–266. https://doi.org/10.1080/02691720903364134

Janis, I. L. (1972). *Victims of groupthink*. Houghton Mifflin.

Keck, S., & Tang, W. (2020). Enhancing the wisdom of the crowd with cognitive-process diversity: The benefits of aggregating intuitive and analytical judgments. *Psychological Science, 31*(10), 1272–1282. https://doi.org/10.2307/4473279210.1177/0956797620941840

Larrick, R. P., Mannes, A. E., & Soll, J. B. (2012). The social psychology of the wisdom of crowds. In J. I. Krueger (Ed.), *Frontiers of social psychology. Social judgment and decision making* (pp. 227–242). Psychology Press. https://doi.org/10.4324/9780203854150

Locke, C. C., & Anderson, C. (2015). The downside of looking like a leader: Power, nonverbal confidence, and participative decision-making. *Journal of Experimental Social Psychology, 58*, 42–47. https://doi.org/10.1016/j.jesp.2014.12.004

Mannes, A. E., Soll, J. B., & Larrick, R. P. (2014). The wisdom of select crowds. *Journal of Personality and Social Psychology, 107*, 276–299. https://doi.org/10.1037/a0036677

Martel, C., Allen, J., Pennycook, G., & Rand, D. G. (2024). Crowds can effectively identify misinformation at scale. *Perspectives in Psychological Science, 19*, 477–488. https://doi.org/10.1177/17456916231190388

Mellers, B., Ungar, L., Baron, J., Ramos, J., Gurcay, B., Fincher, K., et al. (2014). Psychological strategies for winning a geopolitical forecasting tournament. *Psychological Science, 25*(5), 1106–1115. https://doi.org/10.1177/0956797614524255

Mercier, H., & Claidière, N. (2022). Does discussion make crowds any wiser? *Cognition, 222*, 104912. https://doi.org/10.1016/j.cognition.2021.104912

Muchnik, L., Aral, S., & Taylor, S. J. (2013). Social influence bias: A randomized experiment. *Science, 341*, 647–651. https://doi.org/10.1126/science.1240466

Murr, A. E. (2016). The wisdom of crowds: What do citizens forecast for the 2015 British General Election? *Electoral Studies, 41*, 283–288. https://doi.org/10.1016/j.electstud.2015.11.018

Navajas, J., Niella, T., Garbulsky, G., Bahrami, B., & Sigman, M. (2018). Aggregated knowledge from a small number of debates outperforms the wisdom of large crowds. *Nature Human Behaviour, 2,* 126–132. https://doi. org/10.1038/s41562-017-0273-4

O'Leary, D. E. (2017). Crowd performance in prediction of the World Cup 2014. *European Journal of Operational Research, 260,* 715–724. https://doi. org/10.1016/j.ejor.2016.12.043

Page, S. E. (2007). *The difference: How the power of diversity creates better groups, firms, schools, and societies.* Princeton University Press.

Rowe, G., Wright, G., & McColl, A. (2005). Judgment change during Delphi-like procedures: The role of majority influence, expertise, and confidence. *Technological Forecasting and Social Change, 72,* 377–399. https://doi. org/10.1016/j.techfore.2004.03.004

Shi, F., Teplitskiy, M., Duede, E., & Evans, J. A. (2019). The wisdom of polarized crowds. *Nature Human Behaviour, 3*(4), 329–336. https://doi. org/10.1038/s41562-019-0541-6

Simoiu, C., Sumanth, C., Mysore, A., & Goel, S. (2019). Studying the "wisdom of crowds" at scale. In *Seventh AAAI conference on human computation and crowdsourcing (HCOMP-19).*

Staddon, J. (2018, October 7). The devolution of social science. *Quillette.* https://quillette.com/2018/10/07/the-devolution-of-social-science/

Sulik, J., Bahrami, B., & Deroy, O. (2022). The diversity gap: When diversity matters for knowledge. *Perspectives on Psychological Science, 17*(3), 752–767. https://doi.org/10.1177/17456916211006070

Sunstein, C. R. (2006). *Infotopia : How many minds produce knowledge.* Oxford University Press.

Surowiecki, J. (2004). *The wisdom of crowds: Why the many are smarter than the few.* Doubleday.

Toyokawa, W., Whalen, A., & Laland, K. N. (2019). Social learning strategies regulate the wisdom and madness of interactive crowds. *Nature Human Behaviour, 3*(2), 183–193. https://doi.org/10.1038/s41562-018-0518-x

10

Towards a "Wisdom of Scientific Crowds"

The previous chapter examined how research on wisdom of crowds can suggest better processes for determining scientific consensus. However, the limitation of this research is that it involves judgement tasks that differ in content and complexity from those that face scientists. The present chapter looks at decision tasks that could potentially be used to investigate scientific consensus processes—what might be called the "wisdom of scientific crowds".

10.1 Tasks for Studying the Wisdom of Scientific Crowds

To investigate the wisdom of scientific crowds requires suitable judgement tasks. I propose that these tasks should have the following characteristics:

© The Author(s) 2025
A. Jorm, *Expert Consensus in Science*, https://doi.org/10.1007/978-981-97-9222-1_10

1. The task needs to be a realistic one, requiring scientific expertise and a level of complexity typical of scientific judgements.
2. The task must have a verifiable standard against which to judge the quality of consensus decisions. This is a major challenge, as scientific consensus decisions are arguably themselves the validity standard in most realistic cases.
3. The task must be feasible and practical for use within the constraints of a research study. For example, it should be achievable within a reasonable time frame, and the task should not be either too easy or too difficult to come to the correct decision, so that differences between alternative consensus processes are detectable.

Searching the literature for suitable tasks that have these characteristics, I have identified seven candidates. Below I describe these tasks and the main findings concerning the wisdom of scientific crowds when they have been used.

10.1.1 Predicting the Outcomes of Experiments or Clinical Trials

A number of studies have looked at the ability of experts to predict the outcomes of experiments or clinical trials. In these studies, experts are provided with a clear description of the methods involved in the experiments or trials, but the outcomes are unknown at the time of the predictions. Such studies have been carried out with experiments in development economics (DellaVigna & Pope, 2018) and policy interventions (Otis, 2022), and clinical trials in neurology (Atanasov et al., 2022). The accuracy of predictions about individual experiments or trials can be quite variable (Atanasov et al., 2022; DellaVigna & Pope, 2018). To study the wisdom of scientific crowds requires a set of predictions about a large number of experiments, so that broad trends are identifiable.

A good example of such a study is one by Otis (2022), which asked 863 academic experts to provide predictions about social policy interventions that had been tested in seven randomized controlled trials. The trials covered such topics as the effect of monetary incentives versus behavioural interventions on uptake of COVID-19 vaccination in

Sweden, and the effects of cash transfers versus psychotherapy on intimate partner violence and household consumption in women from rural Kenya. A typical expert in this study was a faculty member or PhD student in economics. The experts were presented with pairs of policies and asked to say which one would have the greater effect on a particular outcome. The predictions were made independently by the experts and then the researcher aggregated the predictions by creating random crowds of various sizes. Individual experts correctly chose the better policy intervention 65% of the time, which is better than chance (50%). However, when they were aggregated into crowds of 30 experts, the crowd's mean choice was right 85% of the time, and this increased to 92% when the analysis was restricted to policies with statistically significant differences. Thus, this study showed a clear wisdom-of-scientific-crowds effect in the predictions made.

10.1.2   Predicting the Replicability of Experiments

Due to the concern that many reported findings are not replicable, particularly in psychology and social sciences, there have been major projects to try to reproduce multiple published findings. These projects have found that many published findings cannot the replicated. In a number of these projects, the researchers have also asked experts to estimate the likely replicability of experimental studies using either ratings or prediction markets, and these estimates have been compared with the observed replicability. Such replicability estimates have been made for experiments in the social sciences (Camerer et al., 2018), psychology (Dreber et al., 2015; Forsell et al., 2019), economics (Camerer et al., 2016) and preclinical studies of cancer (Benjamin et al., 2017). In most of these projects, experts were able to predict replicability better than chance (in some studies, considerably more so), although the study of preclinical cancer experiments found that predictions were far too optimistic (Benjamin et al., 2017).

The study by Camerer et al. (2018) illustrates the methods used. They looked at the replicability of 21 experimental studies in social sciences which had been published in the prestigious journals *Nature* or *Science*. They repeated these studies using large samples, but were only able to replicate the findings of 13 of the 21 experiments (62%). To see whether experts could predict replicability, they gave the experts the details of each experiment and asked them to estimate the likelihood that each finding could be replicated. They also set up a prediction market where experts were given 100 tokens which they could use to trade shares on outcomes. For each share they held when the market closed, experts received 1 token if an outcome was realized or 0 otherwise. Both the ratings and the prediction market prices predicted the replicability of experiments quite well, estimating that 61% of the experiments would replicate, compared to the observed value of 62%. When the predicted probabilities of replication were compared to whether experiments replicated, the correlation was very high-- prediction market beliefs correlated 0.84 and survey beliefs 0.76 with successful replication. These results indicate that in this case the scientific crowd was making wise decisions.

10.1.3 Forecasting Infectious Diseases

Forecasting of infectious disease cases can be useful in public health planning and has traditionally been based on historical trends or mathematical models. However, a number of studies have been carried out comparing expert crowd predictions with traditional methods. These have covered monkeypox cases globally (McAndrew et al., 2022), influenza in the United States (Farrow et al., 2017; Polgreen et al., 2007), chikungunya in Central America (Farrow et al., 2017), COVID-19 in Germany and Poland (Bosse et al., 2022) and influenza, enterovirus and dengue fever in Taiwan (Li et al., 2016; Tung et al., 2015). Some studies have collected direct forecasts by experts and aggregated them (Bosse et al., 2022; Farrow et al., 2017; McAndrew et al., 2022), whereas others have set up prediction markets to provide forecasts (Li et al., 2016; Polgreen et al., 2007; Tung et al., 2015). A number of these studies show that expert crowd forecasts can be better than projections from historical data and predictions from statistical models (Bosse et al., 2022; Farrow et al., 2017; Li et al., 2016; Tung et al., 2015). There is also evidence for a wisdom-of-crowds effect, with aggregated crowd results more frequently closer to the true value than any individual expert (Farrow et al., 2017).

A study by Bosse et al. (2022) of forecasting COVID-19 in Germany and Poland illustrates the approach. These researchers compared mathematical models with human forecasts for COVID-19 cases and deaths over the coming two weeks. There were 32 forecasters, who had backgrounds in statistics, forecasting or epidemiology, and around half regarded themselves as "experts" in the area. They were asked to make forecasts weekly on a web application set up for the purpose and then the individual forecasts were combined by calculating means. The researchers found that the crowd consistently outperformed the mathematical models at forecasting cases, but some models did better than the crowd at forecasting deaths.

10.1.4   Predicting the Progression of Immunization Trials

New medical treatments must go through multiple phases of testing before they are approved for clinical use. In the area of infectious diseases, only about 17% of vaccines and treatments that go into Phase I testing are eventually approved. When dealing with rapidly moving pandemics such as COVID-19, it would be useful to be able to predict which ones are most likely to reach the approval stage, so that efforts could be focused on these. Such predictions were investigated by Atanasov et al. (2022) in relation to COVID-19 vaccines and treatments. They compared machine learning predictions with expert predictions on specific questions such as "Will the Coronavirus vaccine by Vaxart Inc. advance from Phase I testing to Phase II by December 31, 2021?" When the human experts were making their predictions, they were provided with computerized data on progress with past treatments that they could interrogate to aid their judgements. The experts in this study were either people with experience in the life sciences, people who had an interest in the topic, or "superforecasters" who had previously participated in forecasting tournaments and found to be in the top 2% in performance. The independent predictions of the experts were aggregated to create a "crowd" using means or medians. The researchers found that the crowd consistently made better predictions than machine learning and concluded that collective human judgements could be practically useful in this area. Although this study did not specifically use experts with relevant scientific knowledge, the method could potentially be used to investigate the wisdom of scientific crowds.

10.1.5 Improving the Accuracy of Medical Diagnoses

Individual doctors sometimes make errors when diagnosing diseases. Several studies have looked at whether aggregating the diagnostic judgements of several doctors can boost accuracy. Improvements to accuracy from crowd aggregation have been found for diagnoses in breast and skin cancer (Kurvers et al., 2015, 2016) and for a range of diagnoses (Barnett et al., 2019).

A study by Kurvers et al. (2015) on skin cancer diagnoses illustrates this area of research. These researchers had 122 medical professionals work via the internet to independently classify images of skin lesions as either malignant or benign using different approaches to assess the lesions. The accuracy of their classifications was judged against histopathologic information. When using pattern analysis (simultaneous assessment of the diagnostic value of all image features), individual medical professionals had a true positive rate of 83% (i.e. malignant lesions were correctly identified 83% of the time) and a false positive rate of 17% (i.e. benign lesions were incorrectly identified as malignant 17% of the time). If the professionals' diagnoses were aggregated into crowds by taking the most common classification given by a set of professionals, the accuracy improved as the size of the crowd increased. For crowds of 11 professionals, for example, the true positive rate increased from 83% to 97% and the false positive rate decreased from 17% to 12%, showing a clear wisdom-of-crowds effect.

10.1.6   Predicting the Citation Potential of Scientific Articles

Journal editors sometimes try to boost the standing of their journal by selecting manuscripts for publication that they believe will be highly cited. A study of the editors of the British medical journal *BMJ* assessed whether editors could predict the citation potential of manuscripts submitted (Schroter et al., 2022). This study asked ten editors to predict the number of citations for the year of publication and the following year. Overall, prediction was poor, both for individual editors and for aggregated judgements. The authors concluded that "there is no wisdom of crowd when it comes to *BMJ* editors" (p. 1). While this result could indicate that the task of prediction is too difficult and not suitable for studying the wisdom of scientific crowds, the *BMJ* is a very prestigious journal which may not attract many poor submissions. The prediction task may be easier if there is a wider range of submission quality. This judgement task may be worth exploring further in journals with a wider range in quality of submissions.

10.1.7   Improving the Sophistication of Scientific Models

The previous examples of wisdom-of-scientific-crowd tasks have all involved making quantitative estimates or binary decisions in situations where there is a verifiable standard of truth to judge performance against. Aminpour et al. (2021) have investigated a very different type of scientific task—producing a causal model which involves complex interconnecting factors and for which there is no simple standard of truth. They asked groups of stakeholders to produce a causal model of a fishery ecosystem, specifically the population dynamics and fisheries management for striped bass in Massachusetts, United States. The researchers used a diverse group of stakeholders, including recreational fishers, commercial fishers and local fisheries managers. The stakeholders worked independently using an online mental modelling technology to draw their own causal models. The individuals' drawings were then mathematically combined into a collective model for the fishery. The researchers were interested in whether a diverse crowd would produce a better-quality model than a homogeneous one, so they developed collective models separately for the different groups of stakeholders, as well as for a crowd involving all stakeholders. To judge the quality of the models, the researchers interviewed a diverse group of eight scientists covering areas like conservation, natural resource management, fisheries biology, economics and social sciences. The scientists were blinded to the identity of the group that produced the aggregated models and were asked to judge the accuracy of the models on a scale from 1 (very inaccurate) to 7 (very accurate). The researchers found that the model produced by the diverse crowd was superior to those produced by homogeneous crowds.

Although this task involved stakeholders rather than scientists and the standard of truth was itself a consensus of experts, it does allow a study of a realistic scientific process that is different from the ones described above. Hence, it merits further consideration for building a knowledge base on the wisdom of scientific crowds.

10.2 The Criteria for a Wise Scientific Crowd

In the previous chapter, I reviewed the wisdom-of-crowds literature and proposed that crowd decisions were likely to be better when the following criteria were met: (1) expertise, (2) cognitive diversity, (3) independence of judgements and (4) opportunity for sharing. The evidence supporting these criteria came from non-scientific judgement tasks. The task for wisdom-of-scientific-crowds research is to find out whether these or other criteria are associated with better scientific crowd decisions. Given the limited amount of research involving scientific crowds to date, only the most tentative conclusions can be drawn. Some research has been carried out on expertise, cognitive diversity and opportunity for sharing, which is reviewed below.

10.2.1 Expertise

A number of studies have looked at whether individual experts make better judgements. However, most of these studies did not aggregate the experts and non-experts into crowds to see whether expert crowds do better.

Benjamin et al. (2017) looked at whether cancer researchers can predict whether preclinical cancer studies can be replicated. They found mixed results on expertise. Experts with more highly cited publications were found to be more accurate, but experts with topic-specific expertise were less accurate because they tended to be overconfident that experiments would replicate. The researchers also did not look at the impact of expertise of crowds.

DellaVigna and Pope (2018) looked at the ability of various groups to predict the results of a complex experiment in behavioural economics. These researchers compared the predictions of academic experts, undergraduate students, MBA students and Amazon Mechanical Turk workers (people with no specific expertise who are paid small amounts to perform online tasks). The academic experts did better than the students or Mechanical Turk workers. However, the researchers found that if they selected out individuals from the group of non-experts who consistently made accurate predictions (so-called superforecasters), these individuals could predict just as well as the academic experts.

Barnett et al. (2019) looked at the effect of expertise on accuracy of medical diagnoses. They found that individual physicians were more accurate than individual medical students. However, when diagnostic judgements were aggregated into crowds, accuracy was much improved and similar for both physician and student crowds.

Atanasov et al. (2022) looked at crowd prediction of responses for three randomized controlled trials in neurology. They found that prediction was poor overall and no better for co-investigators in these trials (who would have had specific expertise) compared to independent experts. These researchers did not look at the wisdom of expert versus non-expert crowds.

Hoogeveen et al. (2020) looked at whether laypeople could predict the replicability of social science experiments and found above chance accuracy. However, their "laypeople" included graduate students, so many had some expertise in social science research. The researchers then compared their findings with earlier studies in which experts predicted the replicability of similar experiments. They found that experts did better than laypeople, but the laypeople could perform in the range of the experts if they were provided with additional information about the strength of the evidence in the original studies. This additional information was arguably raising the laypeople's level of expertise. Again, aggregated crowd expertise was not specifically examined.

Only limited conclusions can be drawn from these studies. In some of the studies, expertise was associated with better performance at the individual level. When non-experts are aggregated into crowds, they may do as well as individual experts. However, little is known about the relative performance of expert versus non-expert crowds.

10.2.2 Cognitive Diversity

Only the study by Aminpour et al. (2021) on creating scientific models has examined the role of cognitive diversity. These researchers asked individual stakeholders to draw a model of a fishery ecosystem. The stakeholders were drawn from recreational fishers, commercial fishers and local fisheries managers. The researchers mathematically combined the individuals' models to create a crowd model. They found that a crowd model based on inputs from a diverse stakeholder crowd was more sophisticated than the ones based on inputs from homogeneous crowds, supporting a role of cognitive diversity in producing better scientific models.

10.2.3 Opportunity for Sharing

Opportunity for sharing has not been directly studied in the wisdom-of-scientific-crowds tasks. However, there are a number of studies comparing the results of prediction markets with surveys, which provide some relevant evidence. While the tasks involve different ways of eliciting judgements, they also differ in whether information is shared. In a prediction market, experts can see the prices paid by other experts, which is an indicator of their predictions, whereas in a survey there is no sharing.

Prediction markets have been compared with surveys in several studies examining whether experiments in various social sciences are replicable. Two studies found that prediction markets did better at predicting replicability than surveys (Dreber et al., 2015; Forsell et al., 2019), while other studies found no difference (Camerer et al., 2016; Camerer et al., 2018). This is clearly a case of "more research is required".

10.3   Conclusion

Based on the range of tasks available, it is feasible to investigate the wisdom of crowd judgements in realistic scientific tasks. Such research would provide a firmer basis for methods for determining deliberative consensus. In the meantime, the principles for quality consensus judgements from other wisdom-of-crowds tasks provide a provisional basis for how to carry out deliberative scientific consensus: expertise, cognitive diversity, independence of judgements and opportunity for sharing.

References

Aminpour, P., Gray, S. A., Singer, A., Scyphers, S. B., Jetter, A. J., Jordan, R., et al. (2021). The diversity bonus in pooling local knowledge about complex problems. *Proceedings of the National Academy of Sciences USA, 118*(5), e2016887118. https://doi.org/10.1073/pnas.2016887118

Atanasov, P. D., Joseph, R., Feijoo, F., Marshall, M., Conway, A., & Siddiqui, S. (2022). Human forest vs. random forest in time-sensitive COVID-19 clinical trial prediction. *SSRN*. https://doi.org/10.2139/ssrn.3981732

Barnett, M. L., Boddupalli, D., Nundy, S., & Bates, D. W. (2019). Comparative accuracy of diagnosis by collective intelligence of multiple physicians vs individual physicians. *JAMA Network Open, 2*(3), e190096. https://doi.org/10.1001/jamanetworkopen.2019.0096

Benjamin, D., Mandel, D. R., & Kimmelman, J. (2017). Can cancer researchers accurately judge whether preclinical reports will reproduce? *PLoS Biology, 15*, e2002212. https://doi.org/10.1371/journal.pbio.2002212

Bosse, N. I., Abbott, S., Bracher, J., Hain, H., Quilty, B. J., Jit, M., et al. (2022). Comparing human and model-based forecasts of COVID-19 in Germany and Poland. *PLoS Computational Biology, 18*(9), e1010405. https://doi.org/10.1371/journal.pcbi.1010405

Camerer, C. F., Dreber, A., Forsell, E., Ho, T. H., Huber, J., Johannesson, M., et al. (2016). Evaluating replicability of laboratory experiments in economics. *Science, 351,* 1433–1436. https://doi.org/10.1126/science.aaf0

Camerer, C. F., Dreber, A., Holzmeister, F., Ho, T. H., Huber, J., Johannesson, M., et al. (2018). Evaluating the replicability of social science experiments in Nature and Science between 2010 and 2015. *Nature Human Behaviour, 2*(9), 637–644. https://doi.org/10.1038/s41562-018-0399-z

DellaVigna, S., & Pope, D. (2018). Predicting experimental results: Who knows what? *Journal of Political Economy, 126,* 2410–2456. https://doi.org/10.1086/699976

Dreber, A., Pfeiffer, T., Almenberg, J., Isaksson, S., Wilson, B., Chen, Y., et al. (2015). Using prediction markets to estimate the reproducibility of scientific research. *PNAS, 112,* 15343–15347. https://doi.org/10.1073/pnas.1516179112

Farrow, D. C., Brooks, L. C., Hyun, S., Tibshirani, R. J., Burke, D. S., & Rosenfeld, R. (2017). A human judgment approach to epidemiological forecasting. *PLoS Computational Biololgy, 13*(3), e1005248. https://doi.org/10.1371/journal.pcbi.1005248

Forsell, E., Viganola, D., Pfeiffer, T., Almenberg, J., Wilson, B., Chen, Y., et al. (2019). Predicting replication outcomes in the Many Labs 2 study. *Journal of Economic Psychology, 75,* 102117. https://doi.org/10.1016/j.joep.2018.10.009

Hoogeveen, S., Sarafoglou, A., & Wagenmakers, E. J. (2020). Laypeople can predict which social-science studies will be replicated successfully. *Advances in Methods and Practices in Psychological Science, 3,* 267–285. https://doi.org/10.1177/251524592091966

Kurvers, R. H., Herzog, S. M., Hertwig, R., Krause, J., Carney, P. A., Bogart, A., et al. (2016). Boosting medical diagnostics by pooling independent judgments. *Proceedings of the National Academy of Sciences USA, 113*(31), 8777–8782. https://doi.org/10.1073/pnas.1601827113

Kurvers, R. H., Krause, J., Argenziano, G., Zalaudek, I., & Wolf, M. (2015). Detection accuracy of collective intelligence assessments for skin cancer diagnosis. *JAMA Dermatology, 151*(12), 1346–1353. https://doi.org/10.1001/jamadermatol.2015.3149

Li, E. Y., Tung, C. Y., & Chang, S. H. (2016). The wisdom of crowds in action: Forecasting epidemic diseases with a web-based prediction market system. *International Journal of Medical Informatics, 92,* 35–43. https://doi.org/10.1016/j.ijmedinf.2016.04.014

McAndrew, T., Majumder, M. S., Lover, A. A., Venkatramanan, S., Bocchini, P., Besiroglu, T., et al. (2022). Early human judgment forecasts of human monkeypox, May 2022. *Lancet Digital Health, 4*(8), e569–e571. https://doi.org/10.1016/S2589-7500(22)00127-3

Otis, N. (2022). Policy choice and the wisdom of crowds. *SSRN.* https://doi.org/10.2139/ssrn.4200841

Polgreen, P. M., Nelson, F. D., & Neumann, G. R. (2007). Use of prediction markets to forecast infectious disease activity. *Clinical Infectious Diseases, 44*(2), 272–279. https://doi.org/10.1086/510427

Schroter, S., Weber, W. E. J., Loder, E., Wilkinson, J., & Kirkham, J. J. (2022). Evaluation of editors' abilities to predict the citation potential of research manuscripts submitted to The BMJ: A cohort study. *BMJ, 379,* e073880. https://doi.org/10.1136/bmj-2022-073880

Tung, C. Y., Chou, T. C., & Lin, J. W. (2015). Using prediction markets of market scoring rule to forecast infectious diseases: A case study in Taiwan. *BMC Public Health, 15,* 766. https://doi.org/10.1186/s12889-015-2121-7

11

Using Expert Consensus to Persuade the Public

There are complex scientific issues that have ramifications for the welfare of humanity, and these require that the general public act in certain ways to gain the benefits of scientific knowledge. Such issues include the role of human activity in climate change, and the safety of genetically modified foods and vaccinations. For the required actions to occur, members of the public must accept the current consensus of scientists working on the topic. In Chap. 3, I argued that the primary evidence on some scientific issues is now so complex that it is generally impossible for a scientist to read and critically evaluate it all. While they may be able to master the primary literature in some specialist sub-area, scientists must generally rely on literature reviews from specialists in other areas in order to gain a complete picture. For members of the public, the task is even more daunting, as they will generally lack the scientific expertise to evaluate any of the primary evidence or even to fully understand literature reviews. Rather, they have to accept statements about the consensus of the relevant scientific experts and act on this.

Much of the information that members of the public get about complex scientific issues comes from news and social media, and these often cite "the scientific consensus". To give a few examples of relevant headlines:

© The Author(s) 2025
A. Jorm, *Expert Consensus in Science*, https://doi.org/10.1007/978-981-97-9222-1_11

"Case closed: 99.9% of scientists agree climate emergency caused by humans" (Watts, 2021)
"AAAS Scientists: Consensus on GMO Safety Firmer Than For Human-Induced Climate Change" (Entine, 2015)
"One scientist can be wrong. But deny the scientific consensus at your peril" (Grimes, 2023)
"Corporate advocacy on Carbon Capture at odds with scientific consensus, InfluenceMap Study Reveals" (Bandyopadhyay, 2023)
"DeSantis aide bucks medical consensus that healthy children should get COVID vaccine" (Mitropoulos, 2022)

On the other hand, the public are also exposed to media reports on sceptics who question whether consensus has any legitimate role in science. Chapter 1 began by quoting a number of prominent lay sceptics. To save the reader going back, below I repeat some excerpts:

> Let's be clear: The work of science has nothing whatever to do with consensus…There is no such thing as consensus science. If it's consensus, it isn't science. If it's science, it isn't consensus. Period. (Crichton, 2003, p. 5)

> We do not say that there is a consensus over the second law of thermodynamics, a consensus that Paris is south of London or that two and two are four. We say that these things are the way things are…Numbers are critical to democracy, but science is not a democracy…Science is a matter of evidence, not what a majority of scientists think (Kay, 2007, paras. 4–9)

> the claim that 99 per cent of scientists believe…as if scientific truth is determined by votes rather than facts." (Tony Abbott quoted by Yaxley (2017, p. 2))

This tension between citing consensus to persuade the public and questioning its legitimacy in science raises the question of how to communicate scientific conclusions to the public. Should "scientific consensus" be used as a means of persuasion or is this a counter-productive approach that will lead to further doubt?

11.1 The Persuasiveness of Scientific Consensus

There is psychological research showing that perceptions of scientific consensus are indeed important to public beliefs about a scientific issue. On the topic of climate change, for example, a meta-analysis of correlational studies showed that perceived scientific consensus is one of the strongest correlates of a member of the public's belief in climate change, with a pooled correlation of 0.35 (Hornsey et al., 2016). This is higher than the correlation of 0.25 between objective knowledge and beliefs, indicating that communicating the scientific consensus to the public might be a better strategy than trying to increase knowledge of the topic.

More importantly, experimental studies show that a single exposure to consensus messaging can lead to positive change. An example is a study by van der Linden (2015) on public beliefs about the safety of childhood vaccination. American adults were invited to participate in an online survey experiment in which they were given an expert-consensus message (e.g. "90% of medical scientists agree that vaccines are safe and that all parents should be required to vaccinate their children") or assigned to a control group where they received no message. The group that received the consensus message had reduced concern about vaccine safety, were less likely to believe in a link between vaccination and autism, and became more likely to support policies requiring people to vaccinate their children. Such findings have been replicated in studies on a range of topics. A meta-analysis of 43 experimental studies about climate change, genetically modified food and vaccination found a strong effect on perceived scientific consensus (standardized mean difference = 0.55) and a much smaller (but still statistically reliable) effect on belief in scientific facts (standardized mean difference = 0.12) (van Stekelenburg et al., 2022).

There is less research on whether these changes from a single exposure last beyond the end of an experiment. Goldberg et al. (2022) found that the effect of consensus messaging on climate change decayed with time, with 40% of the original effect remaining after 26 days. This finding shows that repeated messages about the scientific consensus will be necessary to produce lasting change in beliefs. Fortunately, the changes in beliefs were most durable in people who were doubtful or dismissive of climate change, and they are the main target for consensus messages.

The most important issue is whether consensus messaging produces behaviour change. Most of the experiments have simply looked at the effects on responses to questions on beliefs about a scientific topic, which may not translate into action. However, there is some evidence that changes in behaviour can occur. Bartoš et al. (2022) did a study in Czechia of public misconceptions about doctors' views about COVID vaccines. They found that 90% of Czech doctors trusted the vaccines, but the public believed that only 50% of doctors did. These researchers looked at the effects of informing the public about the consensus of doctors. They found that such information increased the number of people who intended to get vaccinated by 3 percentage points and actual vaccine uptake by 4 percentage points.

While consensus messaging can have positive effects, it is a double-edged sword. The same mechanism can also be used to spread misinformation and produce undesirable effects. This is illustrated by an online experiment by Logemann et al. (2024) on climate change beliefs of members of the US public. These researchers gave misinformation by presenting a poster on the "Oregon Petition Project", which stated that over 31,000 American scientists participated in a petition claiming that humans are not contributing to climate change and contradicting the claim of a 97% consensus among climate scientists that human activity is contributing to global warming. Compared to a control group, people presented with this misinformation had a lower estimation of scientific consensus and less support for public action to mitigate climate change. This finding is relevant to how dissenting statements like the Nongovernmental International Panel on Climate Change and the World Climate Declaration (see Chap. 4, Case Example 4.4) can potentially undermine the consensus of the Intergovernmental Panel on Climate Change.

11.2 Why Do Some People Reject a Scientific Consensus?

The persuasiveness of consensus messages relies on public trust in scientists. If scientists cannot be trusted, then why take any notice of their consensus? There is indeed evidence that trust in scientists is associated with acceptance of a scientific consensus. Pooling the data from 13 studies, Bogert et al. (2024) found that trust in science had a small but reliable correlation of 0.19 with belief in human-induced climate change. Similarly, representative survey data from the public in 12 countries during 2020 found that trust in scientists was associated with support for and compliance with non-pharmacological interventions during the COVID pandemic (Algan et al., 2021).

Lack of trust in scientists is surprisingly common. In 2020, the Wellcome Trust carried out a global survey of over 140,000 people in more than 140 countries asking what they think and feel about science (Wellcome, 2020). When asked to rate how much they trust scientists in their country, only 43% said they trusted them "a lot". There was substantial variation between countries in level of trust. Trust in scientists was highest in Australia and New Zealand (62%), Western Europe (59%) and North America (54%), and lowest in sub-Saharan Africa (19%), South-East Asia (23%) and Central Asia (28%). Interestingly, trust in scientists was found to be higher in 2020 than when the survey was previously conducted in 2018. The percentage trusting scientists "a lot" increased from 34% to 43% over this period, probably because of the increasing public exposure to scientists during the COVID-19 pandemic.

A factor in rejection of scientific consensus is that people are influenced by their broader worldviews in deciding whether a scientist's opinion on an issue is to be trusted. Kahan et al. (2011) have argued that people form a view about a scientific consensus by trying to recall instances of experts they have observed offering an opinion on an issue—what psychological scientists call the "availability heuristic". However, they are more likely to recall examples of experts taking a position that is compatible with their broader worldviews and so tend to overestimate the degree of scientific support for a position they are predisposed to accept. To test the hypothesis that people evaluate expert opinion according to whether it is compatible with their broader values, Kahan et al. (2011) carried out an experiment in which members of the public were presented with information about the fictional author of a book on either climate change, geologic isolation of nuclear waste or laws concerning concealed weapons. The information provided a photo of the author, presented their qualifications and academic appointments, and gave a summary of the position the author took in their book. The participants were then asked to rate how trustworthy and knowledgeable the expert was on the issue. The researchers found that an expert was rated as more trustworthy and knowledgeable when their book espoused a position that was compatible with the person's broader social values, which they classified as either hierarchical-individualistic or egalitarian-communitarian.

Public trust in a scientific consensus can be deliberately undermined by those who oppose a particular course of action for ideological or financial reasons. They can do this by actively promoting minority contrarian viewpoints in a way that suggests that there is less consensus than is the case. Oreskes and Conway (2010) have shown how, in the United States, opponents of government action on smoking, environmental policies and global warming have sought to sow doubt about the strength of the evidence and degree of expert consensus. These doubts are magnified in the public mind when journalists attempt to show balance by presenting both sides of an issue, even when one side has much less scientific merit than the other. Oreskes and Conway have documented how a small group of prominent US scientists who were not actively involved in researching these topics were motivated by their political values and industry funding to become "merchants of doubt" in order to prevent action being taken.

Because social values are inevitably involved in making policy decisions based on scientific findings, it is important that members of the public be involved in the value judgements involved, as this is likely to elicit greater public trust and support for these policies (Kitcher, 2011). Involving the members of the public in this way involves considerable challenges because of the technical nature of the relevant evidence, but there are ways this can be achieved, such as the citizen juries described in Chap. 7.

A final factor in rejection of a scientific consensus is a person's perception of their own knowledge of the issue. Light et al. (2022) have reported a series of studies of the American public's agreement with scientific consensus on a range of issues. These researchers found that those who are most opposed to the scientific consensus have the lowest levels of objective knowledge, but the highest levels of subjective belief in their own knowledge. In other words, they are overconfident in their knowledge.

We can only speculate on how the above findings might apply to some of the prominent laypeople cited in Chap. 1, who confidently reject the consensus of scientists on issues like climate change and vaccine safety. Plausibly, they may reject the consensus of scientists because they do not trust them, basing this mistrust on the clash between their own values and the values inherent in the scientists' consensus, are influenced by contrarian "merchants of doubt", and may overestimate their own understanding of very technical areas.

11.3 Educating About the Role of Consensus in Science

We are faced with some dilemmas in using scientific consensus to persuade the public. While consensus messaging may change beliefs in a positive way, it can also be used to spread misinformation. Consensus messages are also dependent on the public having trust in scientists, but scientists may be mistrusted if they espouse views incompatible with people's broader values.

A possible pathway is to educate people about the important role of consensus in science more generally, rather than focus solely on consensus messages about specific scientific issues. Support for such an approach comes from an experiment by van Stekelenburg et al. (2021) in which they tried to boost the effect of communicating scientific consensus using a two-step communication strategy. In the first step, US adults with false beliefs about genetically modified food and climate change learned about the value of scientific consensus and how to identify it. They were presented with an infographic on "How to figure out whether a claim is true". This explained how it is very difficult for a person to systematically study the evidence on a topic themselves. However, scientists from all over the world have studied the evidence and developed a consensus. A scientific claim can then be evaluated by looking at the consensus among the relevant scientists. After studying the infographic, the participants in the experiment were exposed to consensus messaging in the form of a news article that opposed their beliefs. The article on genetically modified food, for example, came to the following conclusion:

> In 2014 already, a survey showed that there is a scientific consensus on the safety of genetically engineered food. Dr. Cary Funk from the Pew Research Center: "92% of working Ph.D. biomedical scientists said it is as safe to eat genetically engineered [GE] foods as it is to eat non-GE foods."

This two-step strategy was found to be superior to communicating about the level of consensus alone in correcting misperceptions about genetically modified food, but not about climate change. The researchers speculated that this differential effect may be because Americans have lower trust in climate scientists than in biomedical scientists.

These findings come from a single experiment and need replication. However, if confirmed, they may have implications for science education more broadly. Vickers (2023) has proposed a major change in science education in schools, with less emphasis on assessing the original scientific evidence, which is very complex even for the expert, and more on teaching children how to judge when a scientific consensus does or does not exist. He states:

> Instead of putting all of the emphasis on teaching schoolchildren about science, we could put a far greater emphasis on teaching schoolchildren about scientific communities, and in particular the features of those communities that correlate strongly with trustworthy scientific claims. Specifically, schoolchildren could be taught how to competently judge when a solid scientific consensus does/doesn't exist vis-à-vis some claim of interest. Older teenagers would then leave school less able to work a microscope, perhaps, but more able to digest online information relating to scientific community opinion. (Vickers, 2023, p. 236)

This proposal is admittedly sketchy, but if consensus processes are as central to science as this book argues, a greater understanding of the role of consensus is essential at all levels of science education from high school through to specialist postgraduate training.

References

Algan, Y., Cohen, D., Davoine, E., Foucault, M., & Stantcheva, S. (2021). Trust in scientists in times of pandemic: Panel evidence from 12 countries. *Proceedings of the National Academy of Sciences USA, 118*(40), e2108576118. https://doi.org/10.1073/pnas.2108576118

Bandyopadhyay, K. (2023, December 3). Corporate advocacy on carbon capture at odds with scientific consensus, InfluenceMap study reveals. *Times of India.* https://timesofindia.indiatimes.com/science/corporate-advocacy-on-carbon-capture-at-odds-with-scientific-consensus-influencemap-study-reveals/articleshow/105705769.cms

Bartoš, V., Bauer, M., Cahlikova, J., & Chytilova, J. (2022). Communicating doctors' consensus persistently increases COVID-19 vaccinations. *Nature, 606*(7914), 542–549. https://doi.org/10.1038/s41586-022-04805-y

Bogert, J. M., Buczny, J., Harvey, J. A., & Ellers, J. (2024). The effect of trust in science and media use on public belief in anthropogenic climate change: A meta-analysis. *Environmental Communication, 18,* 484–509. https://doi.org/10.1080/17524032.2023.2280749

Crichton, M. (2003). Aliens cause global warming. In California Institute of Technology (Ed.), *Caltech Michelin Lecture.* Pasadena.

Entine, J. (2015, January 29). AAAS scientists: Consensus on GMO safety firmer than for human-induced climate change. *Huffpost.* https://www.huffpost.com/entry/post_b_6572130

Goldberg, M. H., Gustafson, A., van der Linden, S., Rosenthal, S. A., & Leiserowitz, A. (2022). Communicating the scientific consensus on climate change: Diverse audiences and effects over time. *Environment and Behavior, 54*(7–8), 1133–1165. https://doi.org/10.1177/00139165221129539

Grimes, D. R. (2023, August 27). One scientist can be wrong. But deny the scientific consensus at your peril. *The Guardian.* https://www.theguardian.com/commentisfree/2023/aug/27/mi6-richard-dearlove-covid-artificial-intelligence-misunderstands-science

Hornsey, M. J., Harris, E. A., Bain, P. G., & Fielding, K. S. (2016). Meta-analyses of the determinants and outcomes of belief in climate change. *Nature Climate Change, 6*(6), 622–626. https://doi.org/10.1038/nclimate2943

Kahan, D. M., Jenkins-Smith, H., & Braman, D. (2011). Cultural cognition of scientific consensus. *Journal of Risk Research, 14*, 147–174. https://doi.org/10.1080/13669877.2010.511246

Kay, J. (2007, October 10). Science is the pursuit of the truth, not consensus. *Financial Times.* https://www.ft.com/content/c49c8472-767b-11dc-ad83-0000779fd2ac

Kitcher, P. (2011). *Science in a democratic society.* Prometheus Books.

Light, N., Fernbach, P. M., Rabb, N., Geana, M. V., & Sloman, S. A. (2022). Knowledge overconfidence is associated with anti-consensus views on controversial scientific issues. *Science Advances, 8*(29), eabo0038. https://doi.org/10.1126/sciadv.abo0038

Logemann, H. T., Maertens, R., & van der Linden, S. (2024). The reversed gateway belief model – How misinformation impacts perceptions of the scientific consensus on and attitudes towards climate. *PsyArXiv Preprints.* https://doi.org/10.31234/osf.io/3e4kx

Mitropoulos, A. (2022, March 8). DeSantis aide bucks medical consensus that healthy children should get COVID vaccine. *Good Morning America.* https://www.goodmorningamerica.com/news/story/florida-1st-state-advising-giving-covid-19-vaccine-83301047

Oreskes, N., & Conway, E. M. (2010). *Merchants of doubt: How a handful of scientists obscured the truth on issues from tobacco smoke to global warming* (1st U.S. ed.). Bloomsbury Press.

van der Linden, S. L., Leiserowitz, A. A., Feinberg, G. D., & Maibach, E. W. (2015). The scientific consensus on climate change as a gateway belief: Experimental evidence. *PLoS One, 10*(2), e0118489. https://doi.org/10.1371/journal.pone.0118489

van Stekelenburg, A., Schaap, G., Veling, H., & Buijzen, M. (2021). Boosting understanding and identification of scientific consensus can help to correct false beliefs. *Psychological Science, 32*(10), 1549–1565. https://doi.org/10.1177/09567976211007788

van Stekelenburg, A., Schaap, G., Veling, H., & van 't Riet, J., & Buijzen, M. (2022). Scientific-consensus communication about contested science: A preregistered meta-analysis. *Psychological Science, 33*(12), 1989–2008. https://doi.org/10.1177/09567976221083219

Vickers, P. (2023). *Identifying future-proof science.* Oxford University Press.

Watts, J. (2021, October 20). 'Case closed': 99.9% or scientists agree climate emergency caused by humans. *The Guardian*. https://www.theguardian.com/environment/2021/oct/19/case-closed-999-of-scientists-agree-climate-emergency-caused-by-humans

Wellcome. (2020). *Wellcome global monitor: How Covid-19 affected people's lives and their views about science*. https://cms.wellcome.org/sites/default/files/2021-11/Wellcome-Global-Monitor-Covid.pdf

Yaxley, A. (2017, February 19). *Tony Abbott says climate change action is like trying to 'appease the volcano gods'*. https://www.abc.net.au/news/2017-10-10/tony-abbott-says-action-on-climate-change-is-like-killing-goats/9033090

References

Abbasi, K. (2021). The dangers in policy and practice of following the consensus. *BMJ, 375*, n2885. https://doi.org/10.1136/bmj.n2885

Adam, D. (2023). The consensus projects. *Nature, 617*, 452–454.

Algan, Y., Cohen, D., Davoine, E., Foucault, M., & Stantcheva, S. (2021). Trust in scientists in times of pandemic: Panel evidence from 12 countries. *Proceedings of the National Academy of Sciences USA, 118*(40), e2108576118. https://doi.org/10.1073/pnas.2108576118

Allen, L., de Benoist, B., Dary, O., & Hurrell, R. (Eds.). (2006). *Guidelines on food fortification with micronutrients*. World Health Organization and Food and Agriculture Organization of the United Nations. https://www.who.int/publications/i/item/9241594012

Alonso-Coello, P., Oxman, A. D., Moberg, J., Brignardello-Petersen, R., Akl, E. A., Davoli, M., et al. (2016). GRADE Evidence to Decision (EtD) frameworks: A systematic and transparent approach to making well informed healthcare choices. 2: Clinical practice guidelines. *BMJ, 353*, i2089. https://doi.org/10.1136/bmj.i2089

Altman, D. G., & Simera, I. (2016). A history of the evolution of guidelines for reporting medical research: The long road to the EQUATOR Network.

© The Author(s) 2025
A. Jorm, *Expert Consensus in Science*, https://doi.org/10.1007/978-981-97-9222-1

Journal of the Royal Society of Medicine, 109(2), 67–77. https://doi.org/10.1177/0141076815625599

American Academy of Forensic Sciences. (2024). *Academy Standards Board.* Retrieved August 16, 2024, from https://www.aafs.org/academy-standards-board/about-asb

American Psychological Association. (2018). *APA guidelines for psychological practice with boys and men.* American Psychological Association. https://www.apa.org/about/policy/boys-men-practice-guidelines.pdf

Aminpour, P., Gray, S. A., Singer, A., Scyphers, S. B., Jetter, A. J., Jordan, R., et al. (2021). The diversity bonus in pooling local knowledge about complex problems. *Proceedings of the National Academy of Sciences USA, 118*(5), e2016887118. https://doi.org/10.1073/pnas.2016887118

Anderegg, W. R., Prall, J. W., Harold, J., & Schneider, S. H. (2010). Expert credibility in climate change. *Proceedings of the National Academy of Sciences USA, 107*(27), 12107–12109. https://doi.org/10.1073/pnas.1003187107

Andrews, G., Bell, C., Boyce, P., Gale, C., Lampe, L., Marwat, O., et al. (2018). Royal Australian and New Zealand College of Psychiatrists clinical practice guidelines for the treatment of panic disorder, social anxiety disorder and generalised anxiety disorder. *Australian and New Zealand Journal of Psychiatry, 52*, 1109–1172. https://doi.org/10.1177/0004867418799453

Arechar, A. A., Allen, J., Berinsky, A. J., Cole, R., Epstein, Z., Garimella, K., et al. (2023). Understanding and combatting misinformation across 16 countries on six continents. *Nature Human Behaviour, 7*(9), 1502–1513. https://doi.org/10.1038/s41562-023-01641-6

Ashwell, M., Gibson, S., Bellisle, F., Buttriss, J., Drewnowski, A., Fantino, M., et al. (2020). Expert consensus on low-calorie sweeteners: Facts, research gaps and suggested actions. *Nutrition Research Reviews, 33*(1), 145–154. https://doi.org/10.1017/S0954422419000283

Atanasov, P. D., Joseph, R., Feijoo, F., Marshall, M., Conway, A., & Siddiqui, S. (2022). Human forest vs. random forest in time-sensitive COVID-19 clinical trial prediction. *SSRN.* https://doi.org/10.2139/ssrn.3981732

Australian Research Council. (2024). *Overview of ARC funding process.* Retrieved August 14, 2024, from https://www.arc.gov.au/funding-research/peer-review/overview-arc-funding-process#assessment

Bandyopadhyay, K. (2023, December 3). Corporate advocacy on carbon capture at odds with scientific consensus, InfluenceMap study reveals. *Times of India.* https://timesofindia.indiatimes.com/science/corporate-advocacy-on-carbon-capture-at-odds-with-scientific-consensus-influencemap-study-reveals/articleshow/105705769.cms

Barker, T. H., Stone, J. C., Sears, K., Klugar, M., Leonardi-Bee, J., Tufanaru, C., et al. (2023). Revising the JBI quantitative critical appraisal tools to improve their applicability: An overview of methods and the development process. *JBI Evidence Synthesis*, 21(3), 478–493. https://doi.org/10.11124/JBIES-22-00125

Barnett, M. L., Boddupalli, D., Nundy, S., & Bates, D. W. (2019). Comparative accuracy of diagnosis by collective intelligence of multiple physicians vs individual physicians. *JAMA Network Open*, 2(3), e190096. https://doi.org/10.1001/jamanetworkopen.2019.0096

Baron, R. S. (2005). So right it's wrong: Groupthink and the ubiquitous nature of polarized group decision making. *Advances in Experimental Social Psychology*, 37, 219–253. https://doi.org/10.1016/S0065-2601(05)37004-3

Baron, R. M., & Kenny, D. A. (1986). The moderator-mediator variable distinction in social psychological research: Conceptual, strategic, and statistical considerations. *Journal of Personality and Social Psychology*, 51(6), 1173–1182. https://doi.org/10.1037/0022-3514.51.6.1173

Barrio, J. R. (2009). Consensus science and the peer review. *Molecular Imaging and Biology*, 11(5), 293. https://doi.org/10.1007/s11307-009-0233-0

Bartoš, V., Bauer, M., Cahlikova, J., & Chytilova, J. (2022). Communicating doctors' consensus persistently increases COVID-19 vaccinations. *Nature*, 606(7914), 542–549. https://doi.org/10.1038/s41586-022-04805-y

Beatty, J., & Moore, A. (2010). Should we aim for consensus? *Episteme*, 7(3), 198–214. https://doi.org/10.3366/epi.2010.0203

Becker, J., Brackbill, D., & Centola, D. (2017). Network dynamics of social influence in the wisdom of crowds. *Proceedings of the National Academy of Sciences USA*, 114(26), E5070–E5076. https://doi.org/10.1073/pnas.1615978114

Benjamin, D., Mandel, D. R., & Kimmelman, J. (2017). Can cancer researchers accurately judge whether preclinical reports will reproduce? *PLoS Biology*, 15, e2002212. https://doi.org/10.1371/journal.pbio.2002212

Boccardi, M., Bocchetta, M., Apostolova, L. G., Barnes, J., Bartzokis, G., Corbetta, G., et al. (2015). Delphi definition of the EADC-ADNI harmonized protocol for hippocampal segmentation on magnetic resonance. *Alzheimer's & Dementia*, 11(2), 126–138. https://doi.org/10.1016/j.jalz.2014.02.009

Bornmann, L., Mutz, R., & Daniel, H. D. (2010). A reliability-generalization study of journal peer reviews: A multilevel meta-analysis of inter-rater reliability and its determinants. *PLoS One, 5*(12), e14331. https://doi.org/10.1371/journal.pone.0014331

Bosnjak, M., Fiebach, C. J., Mellor, D., Mueller, S., O'Connor, D. B., Oswald, F. L., & Sokol, R. I. (2022). A template for preregistration of quantitative research in psychology: Report of the joint psychological societies preregistration task force. *American Psychologist, 77*(4), 602–615. https://doi.org/10.1037/amp0000879

Bosse, N. I., Abbott, S., Bracher, J., Hain, H., Quilty, B. J., Jit, M., et al. (2022). Comparing human and model-based forecasts of COVID-19 in Germany and Poland. *PLoS Computational Biology, 18*(9), e1010405. https://doi.org/10.1371/journal.pcbi.1010405

Braun, V., & Clarke, V. (2006). Using thematic analysis in psychology. *Qualitative Research in Psychology, 3*, 77–101.

Braun, V., & Clarke, V. (2019). Reflecting on reflexive thematic analysis. *Qualitative Research in Sport, Exercise and Health, 11*(4), 589–597. https://doi.org/10.1080/2159676X.2019.1628806

Briant, J. (2005). Consensus can be wrong. *Institute of Public Affairs Review, 57*(Dec 2005), 35.

Bruggeman, J., Traag, V. A., & Uitermark, J. (2012). Detecting communities through network data. *American Sociological Review, 77*, 1050–1063. https://doi.org/10.1177/0003122412463574

Buettner, D., Nelson, T., & Veenhoven, R. (2020). Ways to greater happiness: A Delphi study. *Journal of Happiness Studies, 21*(8), 2789–2806. https://doi.org/10.1007/s10902-019-00199-3

Butcher, N. J., Monsour, A., Mew, E. J., Chan, A. W., Moher, D., Mayo-Wilson, E., et al. (2022). Guidelines for reporting outcomes in trial reports: The CONSORT-outcomes 2022 extension. *JAMA, 328*(22), 2252–2264. https://doi.org/10.1001/jama.2022.21022

Camerer, C. F., Dreber, A., Forsell, E., Ho, T. H., Huber, J., Johannesson, M., et al. (2016). Evaluating replicability of laboratory experiments in economics. *Science, 351*, 1433–1436. https://doi.org/10.1126/science.aaf0

Camerer, C. F., Dreber, A., Holzmeister, F., Ho, T. H., Huber, J., Johannesson, M., et al. (2018). Evaluating the replicability of social science experiments in Nature and Science between 2010 and 2015. *Nature Human Behaviour, 2*(9), 637–644. https://doi.org/10.1038/s41562-018-0399-z

Campbell, D. T., & Stanley, J. C. (1966). *Experimental and quasi-experimental designs for research.* Rand McNally. https://books.google.com.au/books?id=kVtqAAAAMAAJ

Cardamone-Breen, M. C., Jorm, A. F., Lawrence, K. A., Mackinnon, A. J., & Yap, M. B. H. (2017). The Parenting to Reduce Adolescent Depression and Anxiety Scale: Assessing parental concordance with parenting guidelines for the prevention of adolescent depression and anxiety disorders. *PeerJ, 5,* e3825. https://doi.org/10.7717/peerj.3825

Cardillo, G., Nosotti, M., Scarci, M., Torre, M., Alloisio, M., Benvenuti, M. R., et al. (2022). Air leak and intraoperative bleeding in thoracic surgery: A Delphi consensus among the members of Italian society of thoracic surgery. *Journal of Thoracic Disease, 14*(10), 3842–3853. https://doi.org/10.21037/jtd-22-619

Carpiano, R. M., Callaghan, T., DiResta, R., Brewer, N. T., Clinton, C., Galvani, A. P., et al. (2023). Confronting the evolution and expansion of anti-vaccine activism in the USA in the COVID-19 era. *Lancet, 401*(10380), 967–970. https://doi.org/10.1016/S0140-6736(23)00136-8

Catanzaro, M. (2024). Citation cartels help some mathematicians – And their universities – Climb the rankings. *Science, 383*(6682). https://doi.org/10.1126/science.zcl2s6d

Chalmers, I., Atkinson, P., Fenton, M., Firkins, L., Crowe, S., & Cowan, K. (2013). Tackling treatment uncertainties together: The evolution of the James Lind Initiative, 2003-2013. *Journal of the Royal Society of Medicine, 106*(12), 482–491. https://doi.org/10.1177/0141076813493063

Chan, A. W., Tetzlaff, J. M., Altman, D. G., Laupacis, A., Gotzsche, P. C., Krleza-Jeric, K., et al. (2013). SPIRIT 2013 statement: Defining standard protocol items for clinical trials. *Annals of Internal Medicine, 158*(3), 200–207. https://doi.org/10.7326/0003-4819-158-3-201302050-00583

Chandler, J., Churchill, R., Higgins, J. P. T., Lasserson, T., & Tovey, D. (2017). *Methodological expectations of Campbell Collaboration intervention reviews (MECCIR): Reporting standards.* https://campbellcollaboration.org/meccir.html

Chandler, J., & Hopewell, S. (2013). Cochrane methods – Twenty years experience in developing systematic review methods. *Systematic Reviews, 2,* 76. https://doi.org/10.1186/2046-4053-2-76

Chavalarias, D., Wallach, J. D., Li, A. H., & Ioannidis, J. P. (2016). Evolution of reporting p values in the biomedical literature, 1990–2015. *JAMA, 315*(11), 1141–1148. https://doi.org/10.1001/jama.2016.1952

Chawla, D. S. (2021, November 25). Record number of first-time observers get Hubble telescope time. *Nature.* https://doi.org/10.1038/d41586-021-03538-8

Civillini, M. (2023, February 16). World Bank chief to step down early after climate controversy. *Climate Home News.* https://www.climatechangenews.com/2023/02/16/world-bank-chief-steps-down-climate-controversy/#:~:text=World%20Bank%20president%20David%20Malpass,personal%20views%20on%20climate%20change

Clark, C. J., Jussim, L., Frey, K., Stevens, S. T., Al-Gharbi, M., Aquino, K., et al. (2023). Prosocial motives underlie scientific censorship by scientists: A perspective and research agenda. *Proceedings of the National Academy of Sciences U S A, 120*(48), e2301642120. https://doi.org/10.1073/pnas.2301642120

Clement, S., Jarrett, M., Henderson, C., & Thornicroft, G. (2010). Messages to use in population-level campaigns to reduce mental health-related stigma: Consensus development study. *Epidemiologia e Psichiatria Sociale, 19*(1), 72–79. https://doi.org/10.1017/s1121189x00001627

Climate Intelligence. (2023). *World climate declaration: There is no climate emergency.* CLINTEL. Retrieved February 21, 2023, from https://clintel.org/world-climate-declaration/

Cohen, A. A., Kennedy, B. K., Anglas, U., Bronikowski, A. M., Deelen, J., Dufour, F., et al. (2020). Lack of consensus on an aging biology paradigm? A global survey reveals an agreement to disagree, and the need for an interdisciplinary framework. *Mechanisms of Ageing and Development, 191*, 111316. https://doi.org/10.1016/j.mad.2020.111316

Congressional Research Service. (2021). *Evolving assessments of human and natural contributions to climate change.* Congressional Research Service. https://www.everycrsreport.com/files/2021-08-11_R45086_3741fdd7111e5c017404ff7db8d8f5a90fcbe5a7.pdf

Consensus. (2024). *Consensus: AI search engine for research.* Retrieved August 18, 2024, from https://consensus.app/#

Cook, J., Nuccitelli, D., Green, S. A., Richardson, M., Winkler, B., Painting, R., et al. (2013). Quantifying the consensus on anthropogenic global warming in the scientific literature. *Environmental Research Letters, 8*, 024024.

Cook, J., Oreskes, N., Doran, P. T., Anderegg, W. R. L., Verheggen, B., Maibach, E. W., et al. (2016). Consensus on consensus: A synthesis of consensus estimates on human-caused global warming. *Environmental Research Letters, 11*(4), 048002. https://doi.org/10.1088/1748-9326/11/4/048002

Core Writing Team, Lee, H., & Romero, J. (Eds.). (2023). *IPCC, 2023: Climate change 2023: Synthesis report.* IPCC. https://doi.org/10.59327/IPCC/AR6-9789291691647

Crichton, M. (2003). Aliens cause global warming. In California Institute of Technology (Ed.), *Caltech Michelin Lecture.* Pasadena.

Cristea, I. A., & Ioannidis, J. P. A. (2018). P values in display items are ubiquitous and almost invariably significant: A survey of top science journals. *PLoS One, 13*(5), e0197440. https://doi.org/10.1371/journal.pone.0197440

Croce, M. (2019). On what it takes to be an expert. *The Philosophical Quarterly, 69*(274), 1021. https://doi.org/10.1093/pq/pqy044

Curry, J. (2022, October 5). There's no climate emergency. *BizNews.* https://www.biznews.com/global-citizen/2022/10/05/climate-change-2

Curry, J. A., & Webster, P. J. (2013). Climate change: No consensus on consensus. *CAB Reviews, 8*(001). https://doi.org/10.1079/PAVSNNR20138001

Davies, J. (2017). How voting and consensus created the Diagnostic and Statistical Manual of Mental Disorders (DSM-III). *Anthropology & Medicine, 24*, 32–46. https://doi.org/10.1080/13648470.2016.1226684

Davis-Stober, C. P., Budescu, D. V., Dana, J., & Broomell, S. B. (2014). When is a crowd wise? *Decision, 1*, 79–101. https://doi.org/10.1037/dec0000004

de Oliveira, S., & Nisbett, R. E. (2018). Demographically diverse crowds are typically not much wiser than homogeneous crowds. *Proceedings of the National Academy of Sciences U S A, 115*(9), 2066–2071. https://doi.org/10.1073/pnas.1717632115

DellaVigna, S., & Pope, D. (2018). Predicting experimental results: Who knows what? *Journal of Political Economy, 126*, 2410–2456. https://doi.org/10.1086/699976

Dellsén, F. (2021). Consensus versus unanimity: Which carries more weight? *British Journal for the Philosophy of Science.* https://doi.org/10.1086/718273

Dezecache, G., Dockendorff, M., Ferreiro, D. N., Deroy, O., & Bahrami, B. (2022). Democratic forecast: Small groups predict the future better than individuals and crowds. *Journal of Experimental Psychology: Applied, 28*(3), 525–537. https://doi.org/10.1037/xap0000424

Djulbegovic, B., & Guyatt, G. (2019). Evidence vs consensus in clinical practice guidelines. *JAMA, 322*(8), 725–726. https://doi.org/10.1001/jama.2019.9751

Dornbusch, H. J., Hadjipanayis, A., Del Torso, S., Mercier, J. C., Wyder, C., Schrier, L., et al. (2017). We strongly support childhood immunisation—statement from the European Academy of Paediatrics (EAP). *European*

Journal of Pediatrics, *176*(5), 679–680. https://doi.org/10.1007/s00431-017-2885-0

Douglas, H. E. (2009). *Science, policy, and the value-free ideal.* University of Pittsburgh Press. Table of contents only http://www.loc.gov/catdir/toc/fy1002/2009005463.html

Dreber, A., Pfeiffer, T., Almenberg, J., Isaksson, S., Wilson, B., Chen, Y., et al. (2015). Using prediction markets to estimate the reproducibility of scientific research. *PNAS, 112,* 15343–15347. https://doi.org/10.1073/pnas.1516179112

Duarte, J. L., Crawford, J. T., Stern, C., Haidt, J., Jussim, L., & Tetlock, P. E. (2015). Political diversity will improve social psychological science. *Behavioral and Brain Sciences, 38,* e130. https://doi.org/10.1017/S0140525X14000430

Ecklund, E. H., Johnson, D. R., Scheitle, C. P., Matthews, K. R. W., & Lewis, S. W. (2016). Religion among scientists in international context: A new study of scientists in eight regions. *Socius: Sociological Research for a Dynamic World, 2,* 1–9. https://doi.org/10.1177/2378023116664353

Entine, J. (2015, January 29). AAAS scientists: Consensus on GMO safety firmer than for human-induced climate change. *Huffpost.* https://www.huffpost.com/entry/post_b_6572130

Farrow, D. C., Brooks, L. C., Hyun, S., Tibshirani, R. J., Burke, D. S., & Rosenfeld, R. (2017). A human judgment approach to epidemiological forecasting. *PLoS Computational Biololgy, 13*(3), e1005248. https://doi.org/10.1371/journal.pcbi.1005248

Ferguson, J. H. (1996). The NIH Consensus Development Program. The evolution of guidelines. *International Journal of Technology Assessment in Health Care, 12*(3), 460–474. https://www.ncbi.nlm.nih.gov/pubmed/8840666

Ferguson, C. J. (2023). The American Psychological Association's practice guidelines for men and boys: Are they hurting rather than helping male mental wellness? *New Ideas in Psychology, 68,* 100984. https://doi.org/10.1016/j.newideapsych.2022.100984

Ferri, C. P., Prince, M., Brayne, C., Brodaty, H., Fratiglioni, L., Ganguli, M., et al. (2005). Global prevalence of dementia: A Delphi consensus study. *Lancet, 366*(9503), 2112–2117. https://doi.org/10.1016/s0140-6736(05)67889-0

Ferriman, A. (2007). BMJ readers choose the "sanitary revolution" as greatest medical advance since 1840. *BMJ, 334,* 111. https://doi.org/10.1136/bmj.39097.611806.DB

Fisher, R. A. (1925). *Statistical methods for research workers.* Oliver and Boyd.

Fisher, R. A., & Yates, F. (1938). *Statistical tables for biological, agricultural and medical research*. Oliver and Boyd.

Forsell, E., Viganola, D., Pfeiffer, T., Almenberg, J., Wilson, B., Chen, Y., et al. (2019). Predicting replication outcomes in the Many Labs 2 study. *Journal of Economic Psychology, 75*, 102117. https://doi.org/10.1016/j.joep.2018.10.009

Fragopoulou, A., Grigoriev, Y., Johansson, O., Margaritis, L. H., Morgan, L., Richter, E., & Sage, C. (2010). Scientific panel on electromagnetic field health risks: Consensus points, recommendations, and rationales. *Reviews on Environmental Health, 25*(4), 307–317. https://www.ncbi.nlm.nih.gov/pubmed/21268443

Frey, V., & Van de Rijt, A. (2021). Social influence undermines the wisdom of crowds in sequential decision making. *Management Science, 67*, 4273–4286. https://doi.org/10.1287/mnsc.2020.3713

Fricker, M. (2007). *Epistemic injustice: Power and the ethics of knowing*. Oxford University Press. Table of contents only http://www.loc.gov/catdir/toc/ecip0710/2007003067.html

Frisby, C. L., Redding, R. E., O'Donoghue, W. T., & Lilienfeld, S. O. (Eds.). (2023). *Ideological and political bias in psychology*. Springer.

Galletly, C., Castle, D., Dark, F., Humberstone, V., Jablensky, A., Killackey, E., et al. (2016). Royal Australian and New Zealand College of Psychiatrists clinical practice guidelines for the management of schizophrenia and related disorders. *Australian and New Zealand Journal of Psychiatry, 50*(5), 410–472. https://doi.org/10.1177/0004867416641195

Galton, F. (1907). Vox populi. *Nature, 75*(1949), 450–451. https://doi.org/10.1038/075450a0

Gattrell, W. T., Logullo, P., van Zuuren, E. J., Price, A., Hughes, E. L., Blazey, P., et al. (2024). ACCORD (ACcurate COnsensus Reporting Document): A reporting guideline for consensus methods in biomedicine developed via a modified Delphi. *PLoS Medicine, 21*(1), e1004326. https://doi.org/10.1371/journal.pmed.1004326

Global Climate Intelligence Group. (2022). *World climate declaration: There is no climate emergency*. www.clintel.org

Goldberg, M. H., Gustafson, A., van der Linden, S., Rosenthal, S. A., & Leiserowitz, A. (2022). Communicating the scientific consensus on climate change: Diverse audiences and effects over time. *Environment and Behavior, 54*(7–8), 1133–1165. https://doi.org/10.1177/00139165221129539

Goldman, A. I. (2001). Experts: Which ones should you trust? *Philosophy and Phenomenonological Research, 63*, 85–110. https://doi.org/10.1111/j.1933-1592.2001.tb00093.x

Goldman, A. I. (2021). How can you spot the experts? An essay in social episte-mology. *Royal Institute of Philosophy Supplement, 89*, 85–98. https://doi.org/10.1017/S1358246121000060

Gosmann, N. P., Costa, M. A., Jaeger, M. B., Motta, L. S., Frozi, J., Spanemberg, L., et al. (2021). Selective serotonin reuptake inhibitors, and serotonin and norepinephrine reuptake inhibitors for anxiety, obsessive-compulsive, and stress disorders: A 3-level network meta-analysis. *PLoS Medicine, 18*(6), e1003664. https://doi.org/10.1371/journal.pmed.1003664

Granovskiy, B., Gold, J. M., Sumpter, D. J., & Goldstone, R. L. (2015). Integration of social information by human groups. *Topics in Cognitive Science, 7*(3), 469–493. https://doi.org/10.1111/tops.12150

Grimes, D. R. (2023, August 27). One scientist can be wrong. But deny the scientific consensus at your peril. *The Guardian.* https://www.theguardian.com/commentisfree/2023/aug/27/mi6-richard-dearlove-covid-artificial-intelligence-misunderstands-science

Gürçay, B., Mellers, B. A., & Baron, J. (2015). The power of social influence on estimation accuracy. *Journal of Behavioral Decision Making, 28*, 250–261. https://doi.org/10.1002/bdm.1843

Guyatt, G. H., Oxman, A. D., Vist, G. E., Kunz, R., Falck-Ytter, Y., Alonso-Coello, P., et al. (2008). GRADE: An emerging consensus on rating quality of evidence and strength of recommendations. *BMJ, 336*(7650), 924–926. https://doi.org/10.1136/bmj.39489.470347.AD

Haidt, J., & Jussim, L. (2016). Psychological science and viewpoint diversity. *Observer, 28*(2).

Hanson, R. (1995). Could gambling save science? Encouraging an honest consensus. *Social Epistemology, 9*(1), 3–33. https://doi.org/10.1080/02691729508578768

Harb, S. I., Tao, L., Peláez, S., Boruff, J., Rice, D. B., & Shrier, I. (2021). Methodological options of the nominal group technique for survey item elicitation in health research: A scoping review. *Journal of Clinical Epidemiology, 139*, 140–148. https://doi.org/10.1016/j.jclinepi.2021.08.008

Harris, R. (2021, June 29). Climate explained: How the IPCC reaches scientific consensus on climate change. *The Conversation.* https://theconversation.com/climate-explained-how-the-ipcc-reaches-scientific-consensus-on-climate-change-162600

Hart, L. M., Damiano, S. R., Chittleborough, P., Paxton, S. J., & Jorm, A. F. (2014). Parenting to prevent body dissatisfaction and unhealthy eating patterns in preschool children: A Delphi consensus study. *Body Image, 11*(4), 418–425. https://doi.org/10.1016/j.bodyim.2014.06.010

Harvey, K. J., & Diug, B. O. (2018). The value of food fortification as a public health intervention.M edical Journal of Australia, 208(3), 111–112. https://doi.org/10.5694/mja17.01095

Hilbeck, A. B. R., Defarge, N., Steinbrecher, R., Szekacs, A., Wickson, F., Antoniou, M., Bereano, P. L., Clark, E. A., Hansen, M., Novotony, E., Heinemann, J., Meyer, H., Shiva, V., & Wynne, B. (2015). No scientific consensus on GMO safety. *Environmental Sciences Europe, 27*, 4. https://doi.org/10.1186/s12302-014-0034-1

Hill, A. B. (1965). The environment and disease: Association or causation? *Proceedings of the Royal Society of Medicine, 58*(5), 295–300. https://doi.org/10.1177/003591576505800503

Hoffmann, F., Allers, K., Rombey, T., Helbach, J., Hoffmann, A., Mathes, T., & Pieper, D. (2021). Nearly 80 systematic reviews were published each day: Observational study on trends in epidemiology and reporting over the years 2000–2019. *Journal of Clinical Epidemiology, 138*, 1–11. https://doi.org/10.1016/j.jclinepi.2021.05.022

Hoogeveen, S., Sarafoglou, A., & Wagenmakers, E. J. (2020). Laypeople can predict which social-science studies will be replicated successfully. *Advances in Methods and Practices in Psychological Science, 3*, 267–285. https://doi.org/10.1177/251524592091966

Hornsey, M. J., Harris, E. A., Bain, P. G., & Fielding, K. S. (2016). Meta-analyses of the determinants and outcomes of belief in climate change. *Nature Climate Change, 6*(6), 622–626. https://doi.org/10.1038/nclimate2943

Houghton, J. T., Jenkins, G. J., & Ephraums, J. J. (Eds.). (1990). *Climate change: The IPCC scientific assessment.* Cambridge University Press.

Hsu, C., & Sandford, B. A. (2019). The Delphi technique: Making sense of consensus. *Practical Assessment, Research and Evaluation, 12*, 10. https://doi.org/10.7275/pdz9-th90

Hughes, A., Bonacic, K., Cameron, T., Collins, K., da Costa, F., Debney, A., et al. (2023). Site selection for European native oyster (Ostrea edulis) habitat restoration projects: An expert-derived consensus. *Aquatic Conservation: Marine and Freshwater Ecosystems, 33*, 721–736. https://doi.org/10.1002/aqc.3917

Independent Group of Scientists appointed by the Secretary-General. (2019). *Global sustainable development report 2019: The future is now – Science for achieving sustainable development.* United Nations.

Institute for Ascertaining Scientific Consensus. (2024). *Institute for Ascertaining Scientific Consensus*. Retrieved August 18, 2024, from https://iasc.awh.durham.ac.uk/

Institute of Medicine (U.S.). Committee on Standards for Developing Trustworthy Clinical Practice Guidelines, & Graham, R. (2011). *Clinical practice guidelines we can trust*. National Academies Press.

Intemann, K. (2009). Why diversity matters: Understanding and applying the diversity component of the National Science Foundation's broader impacts criterion. *Social Epistemology, 23*, 249–266. https://doi.org/10.1080/02691720903364134

Intemann, K. (2017). Who needs consensus anyway? Addressing manufactured doubt and increasing public trust in climate science. *Public Affairs Quarterly, 31*(3), 189–208. https://doi.org/10.2307/4473279210.2307/44732792

Intergovernmental Panel on Climate Change. (2023). *Climate change 2023: Synthesis report. Contribution of Working Groups I, II and III to the sixth assessment report of the Intergovernmental Panel on Climate Change*. IPCC.

International Astronomical Union. (2006). *IAU 2006 General Assembly: Result of the IAU resolution votes*. IAU.

International Astronomical Union. (2023). *Working groups*. Retrieved February 22, 2023, from https://www.iau.org/science/scientific_bodies/working_groups/

International Committee of Medical Journal Editors. (2023). *Recommendations for the conduct, reporting, editing, and publication of scholarly work in medical journals*. ICMJE. www.icmje.org

International Organization for Standardization. (2019). In ISO (Ed.), *ISO in brief*. ISO.

Janis, I. L. (1972). *Victims of groupthink*. Houghton Mifflin.

Jarry, J. (2021, April 16). The anti-vaccine propaganda of Robert F. Kennedy, Jr. *Office for Science and Society (OSS) Weekly Newsletter*. https://www.mcgill.ca/oss/article/covid-19-health-pseudoscience/anti-vaccine-propaganda-robert-f-kennedy-jr

JBI. (2013). *JBI levels of evidence*. https://jbi.global/sites/default/files/2019-05/JBI-Levels-of-evidence_2014_0.pdf

Jorm, A. F. (2015). Using the Delphi expert consensus method in mental health research. *Australian and New Zealand Journal of Psychiatry, 49*(10), 887–897. https://doi.org/10.2307/4473279210.1177/0004867415600891

Jureidini, J., & McHenry, L. B. (2022). The illusion of evidence based medicine. *BMJ, 376*, o702. https://doi.org/10.1136/bmj.o702

Kahan, D. M., Jenkins-Smith, H., & Braman, D. (2011). Cultural cognition of scientific consensus. *Journal of Risk Research, 14*, 147–174. https://doi.org/1 0.1080/13669877.2010.511246

Kancherla, V., Botto, L. D., Rowe, L. A., Shlobin, N. A., Caceres, A., Arynchyna-Smith, A., et al. (2022). Preventing birth defects, saving lives, and promoting health equity: An urgent call to action for universal mandatory food fortification with folic acid. *Lancet Global Health, 10*(7), e1053–e1057. https://doi.org/10.1016/S2214-109X(22)00213-3

Kanno, T., Iijima, K., Abe, Y., Koike, T., Shimada, N., Hoshi, T., et al. (2013). Peptic ulcers after the Great East Japan earthquake and tsunami: Possible existence of psychosocial stress ulcers in humans. *Journal of Gastroenterology, 48*(4), 483–490. https://doi.org/10.1007/s00535-012-0681-1

Kay, J. (2007, October 10). Science is the pursuit of the truth, not consensus. *Financial Times.* https://www.ft.com/content/c49c8472-767b-11dc-ad83-0000779fd2ac

Keck, S., & Tang, W. (2020). Enhancing the wisdom of the crowd with cognitive-process diversity: The benefits of aggregating intuitive and analytical judgments. *Psychological Science, 31*(10), 1272–1282. https://doi.org/1 0.2307/4473279210.1177/0956797620941840

Kelso, A. (2016). Review of the Australian code for the responsible conduct of research (2007). *Medical Journal of Australia, 205*(2), 49. https://doi.org/10.5694/mja16.00550

Kennedy-Shaffer, L. (2019). Before p < 0.05 to beyond p < 0.05: Using history to contextualize p-values and significance testing. *American Statistician, 73*(Suppl 1), 82–90. https://doi.org/10.1080/00031305.2018.1537891

Kirshner, R. P. (2013). The accelerating universe: A Nobel surprise. *Proceedings of the American Philosophical Society, 157*, 438–456.

Kitcher, P. (1995). *The advancement of science: Science without legend, objectivity without illusions.* Oxford University Press. https://doi.org/10.1093/0195096533.001.0001

Kitcher, P. (2011). *Science in a democratic society.* Prometheus Books.

Klein, D. B., & Stern, C. (2005). Professors and their politics: The policy views of social scientists. *Critical Review, 17*, 257–303. https://doi.org/10.1080/08913810508443640

Konopka, M. J., Zeegers, M. P., Solberg, P. A., Delhaije, L., Meeusen, R., Ruigrok, G., et al. (2022). Factors associated with high-level endurance performance: An expert consensus derived via the Delphi technique. *PLoS One, 17*(12), e0279492. https://doi.org/10.1371/journal.pone.0279492

Kurvers, R. H., Herzog, S. M., Hertwig, R., Krause, J., Carney, P. A., Bogart, A., et al. (2016). Boosting medical diagnostics by pooling independent judgments. *Proceedings of the National Academy of Sciences USA, 113*(31), 8777–8782. https://doi.org/10.1073/pnas.1601827113

Kurvers, R. H., Krause, J., Argenziano, G., Zalaudek, I., & Wolf, M. (2015). Detection accuracy of collective intelligence assessments for skin cancer diagnosis. *JAMA Dermatology, 151*(12), 1346–1353. https://doi.org/10.1001/jamadermatol.2015.3149

Lancet. (2022). *Information for authors.* www.thelancet.com

Larrick, R. P., Mannes, A. E., & Soll, J. B. (2012). The social psychology of the wisdom of crowds. In J. I. Krueger (Ed.), *Frontiers of social psychology. Social judgment and decision making* (pp. 227–242). Psychology Press. https://doi.org/10.4324/9780203854150

Lazarus, J. V., Romero, D., Kopka, C. J., Karim, S. A., Abu-Raddad, L. J., Almeida, G., et al. (2022). A multinational Delphi consensus to end the COVID-19 public health threat. *Nature, 611*(7935), 332–345. https://doi.org/10.1038/s41586-022-05398-2

Lee, J., & O'Morain, C. (1997). Who should be treated for Helicobacter pylori infection? A review of consensus conferences and guidelines. *Gastroenterology, 113*(6 Suppl), S99–S106. https://doi.org/10.2307/4473279210.1016/s0016-5085(97)80021-2

Levenstein, S., Rosenstock, S., Jacobsen, R. K., & Jorgensen, T. (2015). Psychological stress increases risk for peptic ulcer, regardless of Helicobacter pylori infection or use of nonsteroidal anti-inflammatory drugs. *Clinical Gastroenterology and Hepatology, 13*(3), 498–506 e491. https://doi.org/10.1016/j.cgh.2014.07.052

Li, E. Y., Tung, C. Y., & Chang, S. H. (2016). The wisdom of crowds in action: Forecasting epidemic diseases with a web-based prediction market system. *International Journal of Medical Informatics, 92*, 35–43. https://doi.org/10.1016/j.ijmedinf.2016.04.014

Libby, W. F. (1946). Atmospheric helium three and radiocarbon from cosmic radiation. *Physical Review, 69*, 671–672. https://doi.org/10.1103/PhysRev.69.671.2

Light, N., Fernbach, P. M., Rabb, N., Geana, M. V., & Sloman, S. A. (2022). Knowledge overconfidence is associated with anti-consensus views on controversial scientific issues. *Science Advances, 8*(29), eabo0038. https://doi.org/10.1126/sciadv.abo0038

Lin, A., Andronesi, O., Bogner, W., Choi, I. Y., Coello, E., Cudalbu, C., et al. (2021). Minimum reporting standards for in vivo magnetic resonance spectroscopy (MRSinMRS): Experts' consensus recommendations. *NMR in Biomedicine, 34*(5), e4484. https://doi.org/10.1002/nbm.4484

Liu, X., Cruz Rivera, S., Moher, D., Calvert, M. J., Denniston, A. K., Spirit, A. I., & Group, C.-A. W. (2020). Reporting guidelines for clinical trial reports for interventions involving artificial intelligence: The CONSORT-AI extension. *Lancet Digital Health, 2*(10), e537–e548. https://doi.org/10.1016/S2589-7500(20)30218-1

Locke, C. C., & Anderson, C. (2015). The downside of looking like a leader: Power, nonverbal confidence, and participative decision-making. *Journal of Experimental Social Psychology, 58*, 42–47. https://doi.org/10.1016/j.jesp.2014.12.004

Logemann, H. T., Maertens, R., & van der Linden, S. (2024). The reversed gateway belief model – How misinformation impacts perceptions of the scientific consensus on and attitudes towards climate. *PsyArXiv Preprints*. https://doi.org/10.31234/osf.io/3e4kx

Logg, J. M., & Dorison, C. A. (2021). Pre-registration: Weighing costs and benefits for researchers. *Organizational Behavior and Human Decision Processes, 167*, 18–27. https://doi.org/10.1016/j.obhdp.2021.05.006

Longino, H. E. (2002). *The fate of knowledge*. Princeton University Press.

Longino, H. E. (2004). How values can be good for science. In P. K. M. G. Wolters (Ed.), *Science, values, and objectivity* (pp. 127–142). University of Pittsburgh Press.

Lynas, M., Houlton, B. Z., & Perry, S. (2021). Greater than 99% consensus on human caused climate change in the peer-reviewed scientific literature. *Environmental Research Letters, 16*(11), 114005. https://doi.org/10.2307/4473279210.1088/1748-9326/ac2966

Malow, B. A. (2022). It is time to abolish the clock change and adopt permanent standard time in the United States: A Sleep Research Society position statement. *Sleep, 45*(12), 1–4. https://doi.org/10.1093/sleep/zsac236

Mannes, A. E., Soll, J. B., & Larrick, R. P. (2014). The wisdom of select crowds. *Journal of Personality and Social Psychology, 107*, 276–299. https://doi.org/10.1037/a0036677

Marshall, B. J., & Warren, J. R. (1984). Unidentified curved bacilli in the stomach of patients with gastritis and peptic ulceration. *Lancet, 1*(8390), 1311–1315. https://doi.org/10.2307/4473279210.1016/s0140-6736(84)91816-6

Martel, C., Allen, J., Pennycook, G., & Rand, D. G. (2024). Crowds can effectively identify misinformation at scale. *Perspectives in Psychological Science, 19*, 477–488. https://doi.org/10.1177/17456916231190388

Masci, D. (2009). *Scientists and belief.* https://www.pewresearch.org/religion/2009/11/05/scientists-and-belief/#:~:text=Finally

Maski, K., Trotti, L. M., Kotagal, S., Robert Auger, R., Rowley, J. A., Hashmi, S. D., & Watson, N. F. (2021). Treatment of central disorders of hypersomnolence: An American Academy of Sleep Medicine clinical practice guideline. *Journal of Clinical Sleep Medicine, 17*(9), 1881–1893. https://doi.org/10.5664/jcsm.9328

McAndrew, T., Majumder, M. S., Lover, A. A., Venkatramanan, S., Bocchini, P., Besiroglu, T., et al. (2022). Early human judgment forecasts of human monkeypox, May 2022. *Lancet Digital Health, 4*(8), e569–e571. https://doi.org/10.1016/S2589-7500(22)00127-3

McGlynn, E. A., Kosecoff, J., & Brook, R. H. (1990). Format and conduct of consensus development conferences. Multi-nation comparison. *International Journal of Technology Assessment in Health Care, 6*(3), 450–469. https://doi.org/10.1017/s0266462300001045

Mehand, M. S., Millett, P., Al-Shorbaji, F., Roth, C., Kieny, M. P., & Murgue, B. (2018). World Health Organization methodology to prioritize emerging infectious diseases in need of research and development. *Emerging Infectious Diseases, 24*(9). https://doi.org/10.3201/eid2409.171427

Mellers, B., Ungar, L., Baron, J., Ramos, J., Gurcay, B., Fincher, K., et al. (2014). Psychological strategies for winning a geopolitical forecasting tournament. *Psychological Science, 25*(5), 1106–1115. https://doi.org/10.1177/0956797614524255

Mercier, H., & Claidière, N. (2022). Does discussion make crowds any wiser? *Cognition, 222*, 104912. https://doi.org/10.1016/j.cognition.2021.104912

Miller, B. (2013). When is consensus knowledge based? Distinguishing shared knowledge from mere agreement. *Synthese, 190*, 1293–1316. https://doi.org/10.1007/s11229-012-0225-5

Miller, B. (2019). The social epistemology of consensus and dissent. In M. Fricker, P. J. Graham, D. Henderson, & N. J. L. L. Pedersen (Eds.), *The routledge handbook of social epistemology* (pp. 230–239). Routledge.

Mitropoulos, A. (2022, March 8). DeSantis aide bucks medical consensus that healthy children should get COVID vaccine. *Good Morning America.* https://www.goodmorningamerica.com/news/story/florida-1st-state-advising-giving-covid-19-vaccine-83301047

Moher, D., Galipeau, J., Alam, S., Barbour, V., Bartolomeos, K., Baskin, P., et al. (2017). Core competencies for scientific editors of biomedical journals: Consensus statement. *BMC Medicine, 15*(1), 167. https://doi.org/10.1186/s12916-017-0927-0

Moher, D., Schulz, K. F., Simera, I., & Altman, D. G. (2010). Guidance for developers of health research reporting guidelines. *PLoS Medicine, 7*(2), e1000217. https://doi.org/10.1371/journal.pmed.1000217

Montford, A. (2014). *Fraud, bias and public relations: The 97% 'consensus' and its critics.* https://www.thegwpf.org/content/uploads/2014/09/Warming-consensus-and-it-critics1.pdf

Muchnik, L., Aral, S., & Taylor, S. J. (2013). Social influence bias: A randomized experiment. *Science, 341*, 647–651. https://doi.org/10.1126/science.1240466

Mullis, K., Faloona, F., Scharf, S., Saiki, R., Horn, G., & Erlich, H. (1986). Specific enzymatic amplification of DNA in vitro: The polymerase chain reaction. *Cold Spring Harbour Symposia on Quantitative Biology, 51*(Pt 1), 263–273. https://doi.org/10.1101/sqb.1986.051.01.032

Murr, A. E. (2016). The wisdom of crowds: What do citizens forecast for the 2015 British General Election? *Electoral Studies, 41*, 283–288. https://doi.org/10.1016/j.electstud.2015.11.018

Nakhaie, M. R., & Brym, R. J. (1999). The political attitudes of Canadian professors. *Canadian Journal of Sociology, 24*, 329–353. https://doi.org/10.2307/3341393

National Institutes of Health. (1994). NIH Consensus Conference. Helicobacter pylori in peptic ulcer disease. NIH consensus development panel on helicobacter pylori in peptic ulcer disease. *JAMA, 272*(1), 65–69.

National Institutes of Health. (2023). *NIH consensus development program.* Retrieved February 21, 2023, from https://consensus.nih.gov/

Navajas, J., Niella, T., Garbulsky, G., Bahrami, B., & Sigman, M. (2018). Aggregated knowledge from a small number of debates outperforms the wisdom of large crowds. *Nature Human Behaviour, 2*, 126–132. https://doi.org/10.1038/s41562-017-0273-4

Nejstgaard, C. H., Bero, L., Hrobjartsson, A., Jorgensen, A. W., Jorgensen, K. J., Le, M., & Lundh, A. (2020). Association between conflicts of interest and favourable recommendations in clinical guidelines, advisory committee reports, opinion pieces, and narrative reviews: Systematic review. *BMJ, 371*, m4234. https://doi.org/10.1136/bmj.m4234

NICE National Institute for Health and Care Excellence. (2023). *How we develop NICE guidelines.* https://www.nice.org.uk/about/what-we-do/our-programmes/nice-guidance/nice-guidelines/how-we-develop-nice-guidelines

Nongovernmental International Panel on Climate Change. (2023). *NIPCC Nongovernmental International Panel on Climate Change.* The Heartland Institute. Retrieved February 21, 2023, from http://climatechangereconsidered.org/

Norero, D. (2022). *GMO 25-year safety endorsement: 280 science institutions, more than 3,000 studies.* Genetic Literacy Project. https://geneticliteracyproject.org/2022/01/21/gmo-20-year-safety-endorsement-280-science-institutions-more-3000-studies/

Nosek, B. A., Ebersole, C. R., DeHaven, A. C., & Mellor, D. T. (2018). The preregistration revolution. *Proceedings of the National Academy of Sciences U S A, 115*(11), 2600–2606. https://doi.org/10.1073/pnas.1708274114

O'Connor, D. B., Aggleton, J. P., Chakrabarti, B., Cooper, C. L., Creswell, C., Dunsmuir, S., et al. (2020). Research priorities for the COVID-19 pandemic and beyond: A call to action for psychological science. *British Journal of Psychology, 111*(4), 603–629. https://doi.org/10.1111/bjop.12468

O'Leary, D. E. (2017). Crowd performance in prediction of the World Cup 2014. *European Journal of Operational Research, 260,* 715–724. https://doi.org/10.1016/j.ejor.2016.12.043

OECD/BIPM. (2020). *International regulatory co-operation and international organisations: The case of the International Bureau of Weights and Measures (BIPM).* https://www.oecd.org/en/topics/international-regulatory-co-operation.html

Oreskes, N. (2019). *Why trust science?* Princeton University Press. https://doi.org/10.2307/j.ctv17ppcp4

Oreskes, N., & Conway, E. M. (2010). *Merchants of doubt: How a handful of scientists obscured the truth on issues from tobacco smoke to global warming* (1st U.S. ed.). Bloomsbury Press.

Otis, N. (2022). Policy choice and the wisdom of crowds. *SSRN.* https://doi.org/10.2139/ssrn.4200841

Page, S. E. (2007). *The difference: How the power of diversity creates better groups, firms, schools, and societies.* Princeton University Press.

Page, M. J., McKenzie, J. E., Bossuyt, P. M., Boutron, I., Hoffmann, T. C., Mulrow, C. D., et al. (2021). The PRISMA 2020 statement: An updated guideline for reporting systematic reviews. *BMJ, 372,* n71. https://doi.org/10.1136/bmj.n71

Pashler, H., & Harris, C. R. (2012). Is the replicability crisis overblown? Three arguments examined. *Perspectives in Psychological Science, 7*(6), 531–536. https://doi.org/10.1177/1745691612463401

Paxinos, G., & Watson, C. (1982). *The rat brain in stereotaxic coordinates.* Elsevier.

Perlmutter, S., Aldering, G., Goldhaber, G., Knop, R. A., Nugent, P., Castro, P. G., et al. (1999). Measurements of Ω and Λ from 42 high-redshift supernovae. *Astrophysical Journal, 517*(2), 565. https://doi.org/10.1086/307221

Pezzella, P. (2022). The ICD-11 is now officially in effect. *World Psychiatry, 21*(2), 331–332. https://doi.org/10.1002/wps.20982

Pfeiffer, T., & Almenberg, J. (2010). Prediction markets and their potential role in biomedical research – A review. *Biosystems, 102*(2–3), 71–76. https://doi.org/10.1016/j.biosystems.2010.09.005

Pincock, S. (2005). Nobel Prize winners Robin Warren and Barry Marshall. *Lancet, 366*(9495), 1429. https://doi.org/10.1016/s0140-6736(05)67587-3

Playfor, S., Jenkins, I., Boyles, C., Choonara, I., Davies, G., Haywood, T., et al. (2006). Consensus guidelines on sedation and analgesia in critically ill children. *Intensive Care Medicine, 32*(8), 1125–1136. https://doi.org/10.1007/s00134-006-0190-x

Polgreen, P. M., Nelson, F. D., & Neumann, G. R. (2007). Use of prediction markets to forecast infectious disease activity. *Clinical Infectious Diseases, 44*(2), 272–279. https://doi.org/10.1086/510427

Rescher, N. (1993). *Pluralism: Against the demand for consensus.* Clarendon Press; Oxford University Press. Publisher description http://www.loc.gov/catdir/enhancements/fy0639/93018392-d.html

Resnik, D. B. (2003). Is the precautionary principle unscientific? *Studies in History and Philosophy of Biological and Biomedical Sciences, 34*, 329–344. https://doi.org/10.1016/S1369-8486(02)00074-2

Riess, A. G., Filippenko, A. V., Challis, P., Cloccchiatti, A., Diercks, A., Garnavich, P. M., et al. (1998). Observational evidence from supernovae for an accelerating universe and a cosmological constant. *Astronomical Journal, 116*(3), 1009. https://doi.org/10.1086/300499

RMIT ABC Fact Check. (2020, February 27). Who are the 75 Australian 'scientists and professionals' who say there is no climate emergency? *ABC News.* https://www.abc.net.au/news/2020-02-27/who-are-%2D%2Dscientists-professionals-who-say-no-climate-emergency/11734966

Roberts, J., & Escobar, O. (2015). *Involving communities in deliberation: A study of three citizens' juries on onshore wind farms in Scotland.* https://www.climatexchange.org.uk/wp-content/uploads/2023/09/citizens_juries_report_exec_summary.pdf

Robertson, S., Kremer, P., Aisbett, B., Tran, J., & Cerin, E. (2017). Consensus on measurement properties and feasibility of performance tests for the exercise and sport sciences: A Delphi study. *Sports Medicine - Open, 3*(1), 2. https://doi.org/10.1186/s40798-016-0071-y

Rodhe, H., Charlson, R., & Crawford, E. (1997). Svante Arrhenius and the greenhouse effect. *Ambio, 26,* 2–5.

Ross, A. M., Kelly, C. M., & Jorm, A. F. (2014). Re-development of mental health first aid guidelines for suicidal ideation and behaviour: A Delphi study. *BMC Psychiatry, 14,* 241. https://doi.org/10.1186/s12888-014-0241-8

Rowe, G., Wright, G., & McColl, A. (2005). Judgment change during Delphi-like procedures: The role of majority influence, expertise, and confidence. *Technological Forecasting and Social Change, 72,* 377–399. https://doi.org/10.1016/j.techfore.2004.03.004

Rubin, G., De Wit, N., Meineche-Schmidt, V., Seifert, B., Hall, N., & Hungin, P. (2006). The diagnosis of IBS in primary care: Consensus development using nominal group technique. *Family Practice, 23*(6), 687–692. https://doi.org/10.1093/fampra/cml050

Sackett, D. L., Rosenberg, W. M., Gray, J. A., Haynes, R. B., & Richardson, W. S. (1996). Evidence based medicine: What it is and what it isn't. *BMJ, 312*(7023), 71–72. https://doi.org/10.1136/bmj.312.7023.71

Salomon, J. A. (2010). New disability weights for the global burden of disease. *Bulletin of the World Health Organization, 88*(12), 879. https://doi.org/10.2471/BLT.10.084301

Salomon, J. A., Haagsma, J. A., Davis, A., de Noordhout, C. M., Polinder, S., Havelaar, A. H., et al. (2015). Disability weights for the Global Burden of Disease 2013 study. *Lancet Global Health, 3*(11), e712–e723. https://doi.org/10.1016/S2214-109X(15)00069-8

Sarewitz, D. (2011). The voice of science: Let's agree to disagree. *Nature, 478*(7367), 7. https://doi.org/10.1038/478007a

Schroter, S., Weber, W. E. J., Loder, E., Wilkinson, J., & Kirkham, J. J. (2022). Evaluation of editors' abilities to predict the citation potential of research manuscripts submitted to The BMJ: A cohort study. *BMJ, 379,* e073880. https://doi.org/10.1136/bmj-2022-073880

Seale-Carlisle, T. M., Quigley-McBride, A., Teitcher, J. E. F., Crozier, W. E., Dodson, C. S., & Garrett, B. L. (2024). New insights on expert opinion about eyewitness memory research. *Perspectives in Psychological Science*, 17456916241234837. https://doi.org/10.1177/17456916241234837

Shea, B. J., Grimshaw, J. M., Wells, G. A., Boers, M., Andersson, N., Hamel, C., et al. (2007). Development of AMSTAR: A measurement tool to assess the methodological quality of systematic reviews. *BMC Medical Research Methodology*, 7, 10. https://doi.org/10.1186/1471-2288-7-10

Shi, F., Teplitskiy, M., Duede, E., & Evans, J. A. (2019). The wisdom of polarized crowds. *Nature Human Behaviour*, 3(4), 329–336. https://doi.org/10.1038/s41562-019-0541-6

Shwed, U., & Bearman, P. S. (2010). The temporal structure of scientific consensus formation. *American Sociological Review*, 75(6), 817–840. https://doi.org/10.1177/0003122410388488

Simoiu, C., Sumanth, C., Mysore, A., & Goel, S. (2019). Studying the "wisdom of crowds" at scale. In *Seventh AAAI conference on human computation and crowdsourcing (HCOMP-19)*.

Socol, Y., Shaki, Y. Y., & Yanovskiy, M. (2019). Interests, bias, and consensus in science and regulation. *Dose-Response*, 17(2), 1559325819853669. https://doi.org/10.1177/1559325819853669

Solomon, M. (2001). *Social empiricism*. MIT Press.

Spence, D. (2014). Evidence based medicine is broken. *BMJ*, 348, g22. https://doi.org/10.1136/bmj.g22

Staddon, J. (2018, October 7). The devolution of social science. *Quillette*. https://quillette.com/2018/10/07/the-devolution-of-social-science/

State of Queensland. (2017). *2017 consensus statement: Land use impacts on Great Barrier Reef water quality and ecosystem condition*. State of Queensland.

Stolle, L. B., Nalamasu, R., Pergolizzi, J. V., Jr., Varrassi, G., Magnusson, P., LeQuang, J., et al. (2020). Fact vs fallacy: The anti-vaccine discussion reloaded. *Advances in Therapy*, 37(11), 4481–4490. https://doi.org/10.1007/s12325-020-01502-y

Sturgis, P., Brunton-Smith, I., & Jackson, J. (2021). Trust in science, social consensus and vaccine confidence. *Nature Human Behaviour*, 5(11), 1528–1534. https://doi.org/10.1038/s41562-021-01115-7

Suchy, F. J., Brannon, P. M., Carpenter, T. O., Fernandez, J. R., Gilsanz, V., Gould, J. B., et al. (2010). NIH consensus development conference statement: Lactose intolerance and health. *NIH Consensus and State-of-the-Science Statements*, 27(2), 1–27. https://www.ncbi.nlm.nih.gov/pubmed/20186234

Sulik, J., Bahrami, B., & Deroy, O. (2022). The diversity gap: When diversity matters for knowledge. *Perspectives on Psychological Science*, 17(3), 752–767. https://doi.org/10.1177/17456916211006070

Sunstein, C. R. (2006). *Infotopia : How many minds produce knowledge*. Oxford University Press.

Surowiecki, J. (2004). *The wisdom of crowds: Why the many are smarter than the few*. Doubleday.

Taylor, L. E., Swerdfeger, A. L., & Eslick, G. D. (2014). Vaccines are not associated with autism: An evidence-based meta-analysis of case-control and cohort studies. *Vaccine, 32*(29), 3623–3629. https://doi.org/10.1016/j.vaccine.2014.04.085

Toyokawa, W., Whalen, A., & Laland, K. N. (2019). Social learning strategies regulate the wisdom and madness of interactive crowds. *Nature Human Behaviour, 3*(2), 183–193. https://doi.org/10.1038/s41562-018-0518-x

Tucker, A. (2003). The epistemic signficance of consensus. *Inquiry, 46*, 501–521. https://doi.org/10.1080/00201740310003388

Tung, C. Y., Chou, T. C., & Lin, J. W. (2015). Using prediction markets of market scoring rule to forecast infectious diseases: A case study in Taiwan. *BMC Public Health, 15*, 766. https://doi.org/10.1186/s12889-015-2121-7

Turner, L., Shamseer, L., Altman, D. G., Schulz, K. F., & Moher, D. (2012). Does use of the CONSORT Statement impact the completeness of reporting of randomised controlled trials published in medical journals? A Cochrane review. *Systematic Reviews, 1*, 60. https://doi.org/10.1186/2046-4053-1-60

United Nations Department of Economic and Social Affairs. (2024). *Sustainable development: The 17 goals*. Retrieved August 16, 2024, from https://sdgs.un.org/goals

Vaesen, K., Dusseldorp, G. L., & Brandt, M. J. (2021). An emerging consensus in palaeoanthropology: Demography was the main factor responsible for the disappearance of Neanderthals. *Scientific Reports, 11*(1), 4925. https://doi.org/10.1038/s41598-021-84410-7

van de Werfhorst, H. G. (2020). Are universities left-wing bastions? The political orientation of professors, professionals, and managers in Europe. *British Journal of Sociology, 71*, 47–73. https://doi.org/10.1111/1468-4446.12716

van der Linden, S. L., Leiserowitz, A. A., Feinberg, G. D., & Maibach, E. W. (2015). The scientific consensus on climate change as a gateway belief: Experimental evidence. *PLoS One, 10*(2), e0118489. https://doi.org/10.1371/journal.pone.0118489

van Stekelenburg, A., Schaap, G., Veling, H., & Buijzen, M. (2021). Boosting understanding and identification of scientific consensus can help to correct false beliefs. *Psychological Science, 32*(10), 1549–1565. https://doi.org/10.1177/09567976211007788

van Stekelenburg, A., Schaap, G., Veling, H., & van 't Riet, J., & Buijzen, M. (2022). Scientific-consensus communication about contested science: A preregistered meta-analysis. *Psychological Science, 33*(12), 1989–2008. https://doi.org/10.1177/09567976221083219

Verhagen, A. P., de Vet, H. C., de Bie, R. A., Kessels, A. G., Boers, M., Bouter, L. M., & Knipschild, P. G. (1998). The Delphi list: A criteria list for quality assessment of randomized clinical trials for conducting systematic reviews developed by Delphi consensus. *Journal of Clinical Epidemiology, 51*(12), 1235–1241. https://doi.org/10.1016/s0895-4356(98)00131-0

Verheggen, B., Strengers, B., Cook, J., van Dorland, R., Vringer, K., Peters, J., et al. (2014). Scientists' views about attribution of global warming. *Environmental Science & Technology, 48*(16), 8963–8971. https://doi.org/10.1021/es501998e

Vickers, P. (2023). *Identifying future-proof science*. Oxford University Press.

Voelkel, J., Stagnaro, M., Chu, J., Pink, S. L., Mernyk, J. S., Redekopp, C., et al. (2023). Megastudy identifying effective interventions to strengthen Americans' democratic attitudes. *Science, 386*(6719), eadh4764. https://doi.org/10.1126/science.adh4764

Wang, L., & Poder, T. G. (2023). A systematic review of SF-6D health state valuation studies. *Journal of Medical Economics, 26*(1), 584–593. https://doi.org/10.1080/13696998.2023.2195753

Watts, J. (2021, October 20). 'Case closed': 99.9% or scientists agree climate emergency caused by humans. *The Guardian*. https://www.theguardian.com/environment/2021/oct/19/case-closed-999-of-scientists-agree-climate-emergency-caused-by-humans

Wellcome. (2020). *Wellcome global monitor: How Covid-19 affected people's lives and their views about science*. https://cms.wellcome.org/sites/default/files/2021-11/Wellcome-Global-Monitor-Covid.pdf

Winblad, B., Palmer, K., Kivipelto, M., Jelic, V., Fratiglioni, L., Wahlund, L. O., et al. (2004). Mild cognitive impairment – Beyond controversies, towards a consensus: Report of the International Working Group on Mild Cognitive Impairment. *Journal of Internal Medicine, 256*(3), 240–246. https://doi.org/10.1111/j.1365-2796.2004.01380.x

World Health Organization. (2005, May 23). *Electromagnetic fields project*. WHO. Retrieved 18 Aug 2024 from https://www.who.int/initiatives/the-international-emf-project

Yao, L., Ahmed, M. M., Guyatt, G. H., Yan, P., Hui, X., Wang, Q., et al. (2021). Discordant and inappropriate discordant recommendations in consensus and evidence based guidelines: Empirical analysis. *BMJ, 375*, e066045. https://doi.org/10.1136/bmj-2021-066045

Yap, M. B. H., Cardamone-Breen, M. C., Rapee, R. M., Lawrence, K. A., Mackinnon, A. J., Mahtani, S., & Jorm, A. F. (2019). Medium-term effects of a tailored web-based parenting intervention to reduce adolescent risk of depression and anxiety: 12-month findings from a randomized controlled trial. *Journal of Medical Internet Research, 21*(8), e13628. https://doi.org/10.2196/13628

Yap, M. B., Lawrence, K. A., Rapee, R. M., Cardamone-Breen, M. C., Green, J., & Jorm, A. F. (2017). Partners in parenting: A multi-level web-based approach to support parents in prevention and early intervention for adolescent depression and anxiety. *JMIR Mental Health, 4*(4), e59. https://doi.org/10.2196/mental.8492

Yap, M. B., Pilkington, P. D., Ryan, S. M., & Jorm, A. F. (2014b). Parental factors associated with depression and anxiety in young people: A systematic review and meta-analysis. *Journal of Affective Disorders, 156*, 8–23. https://doi.org/10.1016/j.jad.2013.11.007

Yap, M. B., Pilkington, P. D., Ryan, S. M., Kelly, C. M., & Jorm, A. F. (2014a). Parenting strategies for reducing the risk of adolescent depression and anxiety disorders: A Delphi consensus study. *Journal of Affective Disorders, 156*, 67–75. https://doi.org/10.1016/j.jad.2013.11.017

Yaxley, A. (2017, February 19). *Tony Abbott says climate change action is like trying to 'appease the volcano gods'.* https://www.abc.net.au/news/2017-10-10/tony-abbott-says-action-on-climate-change-is-like-killing-goats/9033090

Zachar, P., & Kendler, K. S. (2012). The removal of Pluto from the class of planets and homosexuality from the class of psychiatric disorders: A comparison. *Philosophy, Ethics, and Humanities in Medicine, 7*, 4. https://doi.org/10.1186/1747-5341-7-4

Index

© The Author(s) 2025
A. Jorm, *Expert Consensus in Science*, https://doi.org/10.1007/978-981-97-9222-1